T0253822

Springer Undergraduate Mathematics Series

Advisory Board

M.A.J. Chaplain, *University of St. Andrews, St. Andrews, Scotland, UK*
A. MacIntyre, *Queen Mary University of London, London, England, UK*
S. Scott, *King's College London, London, England, UK*
N. Snashall, *University of Leicester, England, UK*
E. Süli, *University of Oxford, Oxford, England, UK*
M.R. Tehranchi, *University of Cambridge, Cambridge, England, UK*
J.F. Toland, *University of Cambridge, Cambridge, England, UK*

The Springer Undergraduate Mathematics Series (SUMS) is a series designed for undergraduates in mathematics and the sciences worldwide. From core foundational material to final year topics, SUMS books take a fresh and modern approach. Textual explanations are supported by a wealth of examples, problems and fully-worked solutions, with particular attention paid to universal areas of difficulty. These practical and concise texts are designed for a one- or two-semester course but the self-study approach makes them ideal for independent use.

More information about this series at http://www.springer.com/series/3423

Jean-Baptiste Hiriart-Urruty

Mathematical Tapas

Volume 1 (for Undergraduates)

 Springer

Jean-Baptiste Hiriart-Urruty
Institut de mathématiques
Université Paul Sabatier
Toulouse Cedex 9
France

ISSN 1615-2085   ISSN 2197-4144   (electronic)
Springer Undergraduate Mathematics Series
ISBN 978-3-319-42185-8   ISBN 978-3-319-42186-5   (eBook)
DOI 10.1007/978-3-319-42186-5

Library of Congress Control Number: 2016945771

Mathematics Subject Classification (2010): 00A07, 26Axx, 26Bxx, 26Dxx, 15-XX, 52Axx, 90Cxx

© Springer International Publishing Switzerland 2016
This work is subject to copyright. All rights are reserved by the Publisher, whether the whole or part
of the material is concerned, specifically the rights of translation, reprinting, reuse of illustrations,
recitation, broadcasting, reproduction on microfilms or in any other physical way, and transmission
or information storage and retrieval, electronic adaptation, computer software, or by similar or dissimilar
methodology now known or hereafter developed.
The use of general descriptive names, registered names, trademarks, service marks, etc. in this
publication does not imply, even in the absence of a specific statement, that such names are exempt from
the relevant protective laws and regulations and therefore free for general use.
The publisher, the authors and the editors are safe to assume that the advice and information in this
book are believed to be true and accurate at the date of publication. Neither the publisher nor the
authors or the editors give a warranty, express or implied, with respect to the material contained herein or
for any errors or omissions that may have been made.

Printed on acid-free paper

This Springer imprint is published by Springer Nature
The registered company is Springer International Publishing AG
The registered company address is: Gewerbestrasse 11, 6330 Cham, Switzerland

Foreword

Mathematical tapas... but what are tapas? *Tapas* is a Spanish word (in the Basque country, one would also say *pintxos*) for small savory dishes typically served in bars, with drinks, shared with friends in a relaxed ambiance. The offer is varied, it may be meat, fish, vegetables,... Each guest of the party takes with a stick the tapas he likes best at the moment. This is the spirit of the mathematical tapas that we offer to the reader here.

Before explaining more precisely what these tapas are, let us briefly say what they are not. They are not, at least for the most part, questions extracted from competitions like mathematical olympiads, or national competitions for the best high school students, or other sources of that kind. The mathematical questions posed at these events favour elementary number theory, inequalities on real numbers, and plane geometry (especially properties of triangles). We have, however, selected some tasty tapas of this kind.

The tapas that we offer are mathematical questions to answer, exercises (more than long problems, in spirit). They concern **mathematics at the undergraduate level** (roughly speaking, from the end of high school to the end of the first three years of university);[1] they do not cover the whole spectrum of mathematics, of course. Clearly, they reflect the mathematical interests of the author:

Real analysis, Calculus:
functions of a real variable, multivariate calculus, differential calculus;
Matrices (especially positive semidefinite matrices);
Convexity (sets and functions);
Optimization or "variational" situations...

As a general rule, the topics covered by the book are normally taught at most standard higher education institutions around the world.

How have they been chosen?

- Firstly, because "we like them". In other words, each tapa *shows or proves something*: it could be an interesting inequality among integers, on real numbers or integrals, a useful or surprising property of some mathematical objects, or simply an elegant formula... We are just sensitive to the aesthetics of mathematics.

- Secondly, because they illustrate the following motto: *"if you solve it, you learn something"*. During our career, we taught hundreds of students and, therefore, posed thousands of exercises (in directed sessions of exercise solving, for exams, etc.); but we have not included here (standard) questions whose objective is just to test ability in calculating a derivative, an integral, eigenvalues, etc. We have therefore limited our choice of tapas for this first volume to the (symbolic) number of **333**.

Where have they been chosen from?

I always had a soft spot for the "little" questions of mathematics, the ones that are asked among colleagues at the coffee machine or around a blackboard, in the university dining hall after lunch, and so on. Our mathematical tapas are chosen from among them,

---

[1] "Licence" in the European Higher Education system.

and also from my favorite journals posing such challenges: the *American Mathematical Monthly*, and the French mathematical journals entitled *Revue de la Filière Mathématique* (formerly *Revue de Mathématiques Spéciales*) and *Quadrature*. From time to time, I have posed or solved questions posted in these journals. However, for many tapas, I must confess that I have forgotten their origin or history.

How are they classified?

As in restaurant guides, each tapa has one, two or three stars (★):

- One star (★). Tapas of the first level, for students at the end of high school or in the first year of university.

- Two stars (★★). Tapas of a more advanced level, more appropriate for students in the second year of university. That does not mean that solving them necessarily requires more expertise or wit than the one-starred ones, but sometimes just more mathematical maturity.

- Three stars (★★★). The upper level in the proposed tapas, typically for students in the third year of university. Some may be tough and need more chewing. This category contains our favorite tapas.

How are they presented?

Each tapa begins with a *statement*, of course. The statement may contain the answers to the posed questions; this is the case when the questions or proposals are of the form "Show that... " or "Prove that".

There are no detailed solutions to questions, that would have inflated this book by a factor of three or four. Moreover, in mathematics, there is no uniform and unique way to write down answers. But, to help solve the posed challenges, we have proposed *hints*... They suggest a path to follow. A question without any indication could be impossible or too time-consuming... ; the same question with "spoon-fed" steps could be considered too easy. We have tried to maintain an equilibrium between the two styles. Of course, an interested reader is encouraged to try to chew and swallow the tapas without having recourse to the hints.

When they are not integrated into the statements, we provide *answers* to questions, numerical results for example.

From time to time, we add some *comments*: on the context of the question, on its origin, on a possible extension.

In spite of our efforts, some misprints or even mistakes may have slipped in; we just hope that they are not irreparable.

An essential characteristic of mathematics is to be universal and thus international. So, imagine a student or someone who has some basic knowledge in mathematics in seclusion for some time on an isolated island, or just put into jail... With a pocket book like the one containing these tapas, he might pass an easier time and savour some of them.

Bon appétit !

J.-B. Hiriart-Urruty (JBHU)
Toulouse and the Basque country (academic year 2014–2015)

Notations

All the notations, abbreviations and appellations we use are standard; however we make some of them more precise below.

vs, abbreviation of *versus:* in opposition with, or faced with.

i.e., abbreviation of *id est:* that is to say.

$\mathbb{N}$: the set of natural numbers, that is $\{0, 1, 2...\}$.

$\mathbb{R}^*$: the set of real numbers where the 0 element has been removed.

positive x means $x > 0$; nonnegative x means $x \geqslant 0$. This varies in different countries, sometimes positive is used for nonnegative and strictly positive for positive. Here, we stand by the first appellations.

$f : I \to \mathbb{R}$ is increasing on the interval I means that $f(x) \leqslant f(y)$ whenever $x < y$ in I; to call such functions nondecreasing is a mistake from a logical viewpoint.

$f : I \to \mathbb{R}$ is strictly increasing on the interval I means that $f(x) < f(y)$ whenever $x < y$ in I.

$[a, b]$ denotes the closed interval of $\mathbb{R}$ with end-points a and b.

(a, b) is a somewhat ambiguous notation used to denote the open interval with end-points a and b; in some countries the notation $]a, b[$ is used instead.

log or ln: used indifferently for the natural (or Napierian) logarithm.

$\binom{n}{k} = \frac{n!}{k!(n-k)!}$; also denoted C_n^k in some countries.

$\times$ or, better, $\wedge$, is used for the vector or cross-product of two vectors in $\mathbb{R}^3$.

$\|\cdot\|$: unless specified otherwise, this denotes the usual Euclidean norm in $\mathbb{R}^n$.

tr$A$ or trace(A) stands for the trace of A.

det$A$ stands for the determinant of A.

If $u = (u_1, \dots, u_n)$ and $v = (v_1, \dots, v_n)$ are in $\mathbb{R}^n$, $u^T v = \sum_{i=1}^n u_i v_i$. As a general rule, $\langle \cdot, \cdot \rangle$ denotes a scalar or inner product.

If a function f of n variables is differentiable at $a \in \mathbb{R}^n$, we denote by $\nabla f(a)$ the gradient (vector) of f at a. When $\nabla f(a) = 0$, we say that a is a critical point of f.

$\mathcal{S}_n(\mathbb{R})$ is the set of $n \times n$ real symmetric matrices. Semidefinite and positive definite matrices are always symmetric; this is often recalled and assumed if not.

$A \succcurlyeq 0$ (resp. $A \succ 0$) means that the (symmetric) matrix A is positive semidefinite (resp. positive definite).

F is a primitive or an antiderivative of f on the interval I means that $F' = f$ on I.

l is a cluster point of a sequence (u_n) if it is the limit of some subsequence of (u_n). The set of all cluster points of a sequence is called its limit set.

Classification by levels of difficulty

1–75: ★
76–265: ★★
266–333: ★★★

Classification by topics

Calculus in $\mathbb{R}^n$ (vectors, norms, volumes, surface areas,...): 208, 209, 210, 211, 212, 233, 234, 235, 236, 238, 239, 240, 241, 243, 244, 249, 250, 257, 266, 327
Cauchy–Schwarz inequality: 72, 178, 304, 305
Complex numbers: 65, 66, 99, 103
Convex functions and sets: 57, 128, 133, 134, 135, 136, 137, 196, 207, 208, 223, 226, 265, 288, 298, 313, 320, 321, 322, 325, 330, 333
Differential calculus on functions of a real variable: 43, 44, 45, 46, 47, 50, 51, 53, 54, 55, 56, 59, 60, 119, 121, 123, 124, 125, 126, 127, 130, 131, 132, 140, 141, 142, 165, 214, 215, 245, 287, 293, 294
Curves in the plane: 237, 241, 242, 262, 263
Differential and partial differential equations: 61, 62, 63, 110, 149, 258, 259, 260, 261, 273, 278, 279, 280, 284, 317, 318, 319
Equations: 14, 22, 52, 67, 84, 85
Functional equations: 30, 106, 107, 108, 138, 139, 276, 277, 289
Functions of a real variable: 6, 23, 25, 26, 28, 29, 33, 34, 35, 39, 40, 41, 42, 43, 81, 89, 102, 104, 105, 111, 112, 119, 225, 282, 283, 290
Functions of several variables: 74, 147, 206, 213, 216, 225, 227, 232, 245, 324
Geometry (triangles, ellipses, parabolas, ovals, tetrahedra,...): 75, 152, 153, 154, 155, 231, 237, 246, 247, 248, 249, 250, 262, 263, 315, 328, 329
Groups, rings: 64, 150, 151, 157, 295
Harmonic numbers: 77, 78, 79, 80
Inequalities on real numbers: 1, 2, 3, 4, 5, 7, 16, 36, 76, 83, 84, 86, 209
Integers: 8, 9, 10, 11, 15, 17, 19, 268, 269
Integrals of functions of real variables: 31, 34, 49, 113, 114, 116, 117, 122, 128, 144, 146, 245, 285, 291, 292, 326
Linear algebra: 68, 143, 158, 159, 161, 224, 302
Matrices: 69, 70, 71, 73, 148, 157, 160, 161, 162, 163, 164, 165, 166, 167, 168, 169, 170, 171, 172, 173, 174, 189, 190, 191, 192, 199, 200, 201 ,202, 203, 204, 205, 206, 207, 222, 295, 296, 297, 298, 299, 300, 301, 302, 307, 309, 310, 313, 323
Optimization: 48, 58, 74, 145, 153, 197, 198, 217, 218, 219, 220, 228, 229, 230, 247, 248, 312, 332, 333
Polynomials: 33, 52, 67, 115, 118, 129, 156, 163, 213, 214, 215, 281, 286, 314, 316
Positive definite and positive semidefinite matrices: 73, 173, 179, 180, 181, 182, 184, 185, 186, 187, 188, 303, 308, 311

ix

Chapter 1. Proposals

"Obvious is the most dangerous word in mathematics"
E.T. BELL (1883–1960)

1. ★ *Product of mean values vs mean value of products*

   Consider ordered real numbers $x_1 \leqslant x_2 \leqslant \ldots \leqslant x_n$ and $y_1 \leqslant y_2 \leqslant \ldots \leqslant y_n$.

   1°) Let $\sigma : \{1, 2, \ldots, n\} \to \{1, 2, \ldots, n\}$ be a permutation on the integers $1, 2, \ldots, n$. Show that

   $$\sum_{k=1}^{n} x_k y_{\sigma(k)} \leqslant \sum_{k=1}^{n} x_k y_k. \tag{1}$$

   2°) An application of (1). Prove that

   $$\left(\frac{1}{n} \sum_{k=1}^{n} x_k \right) \left(\frac{1}{n} \sum_{k=1}^{n} y_k \right) \leqslant \left(\frac{1}{n} \sum_{k=1}^{n} x_k y_k \right). \tag{2}$$

---

2. ★ We start with ordered real numbers $x_1 \leqslant x_2 \leqslant \ldots \leqslant x_n$ and $y_1 \leqslant y_2 \leqslant \ldots \leqslant y_n$. Consider now an arbitrary permutation $z_1, z_2, \ldots, z_n$ of $y_1, y_2, \ldots, y_n$. Prove that

   $$\sum_{k=1}^{n} (x_k - y_k)^2 \leqslant \sum_{k=1}^{n} (x_k - z_k)^2.$$

---

3. ★ Consider positive real numbers $x_1, x_2, \ldots, x_n$ satisfying $x_1 + x_2 + \ldots + x_n = 1$. Show that

   $$\sum_{k=1}^{n} \frac{1}{x_k} \geqslant n^2.$$

   Can this lower bound n^2 be attained?

---

4. ★ Consider positive real numbers $a_1, a_2, \ldots, a_n$.

   1°) Prove that the minimal value of $\sum_{i=1}^{n} a_i x_i^2$ over all the nonnegative $x_1, x_2, \ldots x_n$ summing up to 1 is equal to $\mu = 1 / \left(\sum_{i=1}^{n} \frac{1}{a_i} \right)$.

   2°) For which x_i do we have $\sum_{i=1}^{n} a_i x_i^2 = \mu$?

1

© Springer International Publishing Switzerland 2016
J.-B. Hiriart-Urruty, *Mathematical Tapas*, Springer Undergraduate
Mathematics Series, DOI 10.1007/978-3-319-42186-5_1

5. ★ Consider distinct positive real numbers $x_1, x_2, ..., x_n$.

1°) Find the maximum value of

$$x_1 x_{\sigma(1)} + x_2 x_{\sigma(2)} + ... + x_n x_{\sigma(n)}$$

over all permutations σ of $\{1, 2, ... , n\}$.

2°) Determine the permutations (called maximizers) for which the maximal value is achieved.

---

6. ★ Consider positive real numbers x and y satisfying $x + y = 1$. Show that

$$x \log(x) + y \log(y) \geqslant - \log(2). \tag{1}$$

1°) When does equality hold true in (1)?

2°) Deduce from the above that the minimal value of $x^x (1-x)^{1-x}$ over all the x lying in the interval $(0, 1)$ is attained at $x = \frac{1}{2}$.

3°) Generalize to the case of positive real numbers $x_1, x_2, ..., x_n$ satisfying $x_1 + x_2 + ... + x_n = 1$.

---

7. ★ *$(n!)^2$ grows faster than n^n*

Show that

$$(n!)^2 > n^n \text{ for all integers } n \geqslant 3.$$

---

8. ★ *Summing odd and even binomial terms*

By expanding $(1 + 1)^n$ with the help of NEWTON's binomial formula, we know that

$$\sum_{k=0}^{n} \binom{n}{k} = 2^n.$$

Now, what about $S_o = \sum_{\substack{k=0 \\ k \text{ odd}}}^{n} \binom{n}{k}$ and $S_e = \sum_{\substack{k=0 \\ k \text{ even}}}^{n} \binom{n}{k}$?

---

9. ★ *Sums and quotients of successive odd integers*

Let

$$S_n = 1 + 3 + 5 + ... + (2n - 1)$$

denote the sum of the n first odd integers.

1°) Observing that S_n is also $(2n - 1) + ... + 5 + 3 + 1$, determine $2S_n$ and deduce an expression for S_n, in terms of n, in a simple closed form.

$2°)$ An application. Let

$$Q_n = \frac{1 + 3 + ... (2n - 1)}{(2n + 1) + (2n + 3) + ... (4n - 1)},$$

that is, the quotient of the sum of the n first odd integers by the sum of the next n odd integers.

Prove that $Q_n = 1/3$ for all $n \geqslant 1$.

---

10. ★ *Sums of powers vs powers of sums*

Show that

$$\left(\sum_{k=1}^{n} k \right)^2 = \sum_{k=1}^{n} k^3.$$

From this, find

$$\lim_{n \to +\infty} \frac{1^3 + 2^3 + ... + n^3}{n^4}. \tag{1}$$

---

11. ★ *Evaluating sums of powers*

Let $n \geqslant 1$ be a fixed integer. For positive integers k, let

$$S_n(k) = 1^k + 2^k + ... + n^k.$$

$1°)$ What are $S_n(1), S_n(2), S_n(3)$ in terms of n?

$2°)$ Establish the following relationship:

$$\sum_{i=1}^{k} \binom{k+1}{i} S_n(i) = (n+1)^{k+1} - n - 1. \tag{1}$$

$3°)$ Deduce from the above $S_n(5)$ and $S_n(7)$ in terms of n.

---

12. ★ *Constructing an irrational number between two rational numbers*

Let x and y be two different rational numbers, with $x < y$, for example. We want to construct an irrational number z lying between x and y.

Let

$$z = \frac{1 + y\sqrt{2}}{1 + x\sqrt{2}}.$$

Knowing that $\sqrt{2}$ is irrational, show that the z proposed above answers our question.

3

13. ★ *When is $\sqrt{n}$ irrational ?*

Let n be a positive integer.

Show that $\sqrt{n}$ is irrational if and only if n is not a perfect square (that is, the square of an integer).

---

14. ★ *A student in a hurry*

A student goes on his bike from his home to the university. He covers the distance with an average velocity of 10 m/h (ten miles per hour). He intends to go back home again on his bike but faster.

What should his average velocity be on the way back so that his average velocity over the entire journey (round-trip) is 20 m/h?

---

15. ★ *A relation between greatest common divisors*

Here, the greatest common divisor of positive integers a and b is denoted by $a \wedge b$.

Consider an integer $n \geqslant 2$. Show that, for positive integers p and q, we have:

$$(n^p - 1) \wedge (n^q - 1) = n^{p \wedge q} - 1.$$

---

16. ★ *Getting upper bounds for the real numbers $\sqrt[n]{n}$ when $m \geqslant n$*

By studying the variations of the function $x > 0 \mapsto f(x) = x^{\frac{1}{x}}$, determine the largest element in the set $\{f(n) \mid n \text{ positive integer}\}$.

Application. Show that, for positive integers m and n, either $\sqrt[n]{n}$ or $\sqrt[m]{m}$ is bounded above by $\sqrt[3]{3}$. —

---

17. ★ *A primality test*

Let $n \geqslant 2$ be an integer. We denote by $\sigma(n)$ the sum of all divisors of n.

1°) What is $\sigma(n)$ if n is a prime number?

2°) Show that
$$(n \text{ is not prime}) \Rightarrow \left(\sigma(n) \geqslant n + \sqrt{n}\right).$$

3°) Deduce from the above results the following equivalence:

$$(n \text{ is prime}) \Leftrightarrow \left(\sigma(n) < n + \sqrt{n}\right).$$

---

18. ★ *Extracting monotone subsequences from convergent sequences*

Let (u_n) be a sequence of real numbers converging to l.

Show that one can extract from (u_n) a monotone subsequence converging to l.

Show that the extracted subsequence can be chosen to be strictly monotone if (u_n) is not a stationary sequence.

---

19. ★ *The factorial does not act like the exponential*

We know that $\exp(n + m) = \exp(n)\exp(m)$. Consider now the factorial function $n!$ (for nonnegative integers n). By convention, $0! = 1$.

Question: when does $(n + m)!$ equal $n!m!$?

---

20. ★ Let (u_n) be a sequence of nonnegative real numbers such that the corresponding series is convergent, *i.e.* $\sum_{n=1}^{+\infty} u_n < +\infty$. As a consequence, the sequence (u_n) converges to 0.

We assume now that the sequence (u_n) is decreasing.

Prove then that (u_n) converges to 0 faster than $\frac{1}{n}$, that is to say: nu_n converges to 0.

---

21. ★ Let p be a positive integer.

1°) Show that

$$\sum_{n=1}^{+\infty} \frac{1}{n(n+1)(n+2)...(n+p)} = \frac{1}{p.p!}. \tag{1}$$

2°) Application. Determine

$$S_p = \sum_{n=1}^{+\infty} \frac{1}{\left(\dfrac{n+p}{p+1}\right)}. \tag{2}$$

---

22. ★ *Lines or curves in the plane avoiding all the points with rational coordinates*

The set $\mathbb{Q}$ of rational numbers is dense in the whole real line $\mathbb{R}$. In a similar way, the set $\mathbb{Q} \times \mathbb{Q}$ of all points with rational coordinates, *i.e.* $\mathbb{Q} \times \mathbb{Q}$, is dense in the plane $\mathbb{R}^2$.

1°) A straight line in the plane avoiding all the points with rational coordinates.

Question: can the graph of a linear function f, *i.e.* of the form $f(x) = ax$ (with $a \neq 0$), avoid all the non-null points in $\mathbb{Q} \times \mathbb{Q}$?

5

2°) Pieces of curves in the plane avoiding all the points with rational coordinates.

For an integer $n \geqslant 2$, consider the arc

$$\mathcal{C}_n = \left\{ (u, v) \in \mathbb{R}^2 \; ; \; u > 0, v > 0 \text{ and } u^n + v^n = 1 \right\}.$$

Question: does $\mathcal{C}_n$ avoid all the points with positive coordinates in $\mathbb{Q} \times \mathbb{Q}$?

---

23. ★ Let $f : [0, 1] \to \mathbb{R}$ satisfy the following properties:

$$
\begin{aligned}
|f(x) - f(y)| \;&<\; |x - y| \text{ for all } x \neq y \text{ in } [0, 1]; \\
f(0) \;&=\; f(1).
\end{aligned}
$$

Prove that

$$|f(x) - f(y)| < \frac{1}{2} \text{ for all } x \text{ and } y \text{ in } [0, 1].$$

---

24. ★ *Characterizing the convergence of successive approximations*

Let $f : [0, 1] \to [0, 1]$ be continuous. Define the sequence (x_n) by successive approximations as follows:

$$
\begin{cases}
x_0 \in [0, 1]; \\
x_{n+1} = f(x_n) \text{ for nonnegative integers } n.
\end{cases}
$$

Show that (x_n) converges (towards a fixed point of f) if and only if $(x_{n+1} - x_n)$ converges to 0.

---

25. ★ *Increasing functions admit fixed points*

Let $f : [0, 1] \to [0, 1]$ be just increasing.

Prove that there exists a fixed point of f, that is to say, a point x_0 in the interval $[0, 1]$ for which $f(x_0) = x_0$.

---

26. ★ Let $f : [0, 1] \to \mathbb{R}$ be a continuous function satisfying

$$\int_0^1 f(x) \, dx = \frac{1}{2}.$$

Show that there exists an x_0 in the interval $(0, 1)$ such that $f(x_0) = x_0$.

---

27. ★ What is the closure of the set

$$S = \{ \sqrt{m} - \sqrt{n} : m \text{ and } n \text{ are nonnegative integers} \}?$$

---

28. ★ *A continuous function meeting every affine function infinitely often*

Does there exist a continuous function $f : \mathbb{R} \to \mathbb{R}$ whose graph intersects every non-vertical line (*i.e.*, the graph of an affine function $x \in \mathbb{R} \mapsto ax + b$) in infinitely many points?

---

29. ★ *Continuity of the record function*

Consider a function $f : [a, +\infty) \to \mathbb{R}$ and its "record function"

$$x \in [a, +\infty) \mapsto f^*(x) = \max_{u \in [a,x]} |f(u)| .$$

Show that f^* is continuous whenever f is continuous.

---

30. ★ *Functions satisfying $f(x^2) = f(x)$ for all x*

Let $f : \mathbb{R} \to \mathbb{R}$ be a continuous function for which:

$$f(x^2) = f(x) \text{ for all } x \in \mathbb{R}. \tag{1}$$

What can be said about such functions?

---

31. ★ *Special definite integrals involving the sine function*

Let $f : [0, 1] \to \mathbb{R}$ be a continuous function. Show that

$$\int_0^\pi x f(\sin x) \, dx = \pi \int_0^{\frac{\pi}{2}} f(\sin x) \, dx.$$

---

32. ★ *Convergence to the maximal value*

1°) Let $x_1, x_2, \dots, x_n$ be nonnegative real numbers and let $p > 0$. We set $m_p = \left(\sum_{i=1}^n x_i^p \right)^{\frac{1}{p}}$.

Show that m_p converges to $\max_i(x_i)$ as $p \to +\infty$.

2°) In the same vein, consider a continuous function $f : [a, b] \to \mathbb{R}^+$ and let $M_p = \left(\int_a^b [f(x)]^p \, dx \right)^{\frac{1}{p}}$.

Prove that M_p converges to $M = \max_{x \in [a,b]} f(x)$ as $p \to +\infty$.

---

33. ★ *An explicit sequence of polynomial functions converging towards the square root function*

Let $(P_n : \mathbb{R} \to \mathbb{R})_n$ be the sequence of polynomial functions defined recursively as follows:

$$P_0 = 0;$$
$$\text{for all } n \geqslant 0, \ P_{n+1}(x) = \frac{1}{2}\left[x + 2P_n(x) - P_n^2(x)\right].$$

1°) (a) Show that

$$0 \leqslant \sqrt{x} - P_n(x) \leqslant \frac{2\sqrt{x}}{2 + n\sqrt{x}} \text{ for all } x \in [0, 1]. \tag{1}$$

(b) Deduce from (1) that the sequence (P_n) converges towards the square root function $\sqrt{\cdot}$ uniformly on $[0, 1]$.

2°) Prove that, for every $\varepsilon > 0$, there exists a polynomial function Q_ε such that

$$|Q_\varepsilon(x) - |x|| \leqslant \varepsilon \text{ for all } x \in [-1, 1].$$

---

34. ★ Let $f : [0, +\infty) \to [0, +\infty)$ be a continuous function for which there exists a $k \in \mathbb{R}$ such that:

$$f(x) \leqslant k \int_0^x f(t)\, dt \text{ for all } x \geqslant 0.$$

Show that f is necessarily the zero function.

---

35. ★ *A continuous sum of increasing functions*

Let f and $g : \mathbb{R} \to \mathbb{R}$ be increasing functions. We suppose that their sum $f + g$ is continuous.

Are f and g continuous?

---

36. ★ *From $a^2 - b^2$ to $\sin^2(a) - \sin^2(b)$*

Knowing that $a^2 - b^2 = (a + b)(a - b)$ for all real numbers a and b, an inattentive student is tempted to write:

$$\sin^2(a) - \sin^2(b) = \sin(a + b)\sin(a - b). \tag{1}$$

Is the formula above right or not?

---

37. ★ *Four formulas involving the* arctan *function*

1°) Prove that

$$\arctan \frac{1}{2} + \arctan \frac{1}{3} = \frac{\pi}{4};$$ (1-1)

$$\arctan 1 + \arctan 2 + \arctan 3 = \pi.$$ (1-2)

2°) Prove that

$$4 \arctan \frac{1}{5} - \arctan \frac{1}{239} = \frac{\pi}{4}.$$ (2)

3°) Prove that, for all $x \in \mathbb{R}$,

$$\arctan (x + 1) - \arctan (x) = \arctan \left(\frac{1}{1 + x + x^2} \right).$$ (3)

---

38. ★ *Playing with* arcsin *and* arccos *values*

1°) Show that, for x in $[0, 1]$:

$$\arcsin (\sqrt{x}) = \frac{\pi}{4} + \frac{1}{2}\arcsin (2x - 1).$$ (1)

2°) Show that

$$f(x) = \int_0^{\sin^2 x} \arcsin (\sqrt{t}) \, dt + \int_0^{\cos^2 x} \arccos (\sqrt{t}) \, dt$$ (2)

is constant on $\mathbb{R}$ and determine this constant value.

---

39. ★ *An unusual chain rule*

Let $u : I \to (0, +\infty)$ and $v : I \to \mathbb{R}$ be differentiable functions on the interval I. Consider the function f defined on I by $f(x) = u(x)^{v(x)}$.

Show that f is differentiable on I and calculate its derivative.

---

40. ★ *The* LAMBERT *function*

Let $f : \mathbb{R} \to \mathbb{R}$ be defined by $f(x) = x \exp(x)$.

1°) Show that f is a strictly increasing bijection from $[-1, +\infty)$ onto $\left[-\frac{1}{e}, +\infty\right)$. We denote by $g : \left[-\frac{1}{e}, +\infty\right) \to [-1, +\infty)$ the inverse bijection of f.

2°) Show that g is differentiable on $(0, +\infty)$ and that, for $y \in (0, +\infty)$,

$$\begin{cases} g'(y) = \frac{x}{(1+x)y}, \text{ where } x \text{ is the unique} \\ \text{solution of the equation } x \exp(x) = y. \end{cases}$$ (1)

41. ★ *The* GUDERMANN *function*

Let f_1, f_2, f_3, f_4, f_5 be five functions from $I = \left(-\frac{\pi}{2}, \frac{\pi}{2}\right)$ into $\mathbb{R}$ defined as follows:

$$f_1(x) = \operatorname{arcsinh}(\tan x); \quad f_2(x) = \operatorname{arctanh}(\sin x); \tag{1-1}$$

$$f_3(x) = 2\operatorname{arctanh}\left(\tan \frac{x}{2}\right); \quad f_4(x) = \ln\left[\tan\left(\frac{x}{2} + \frac{\pi}{4}\right)\right]; \tag{1-2}$$

$$f_5(x) = \ln\left(\frac{1 + \sin x}{\cos x}\right). \tag{1-3}$$

1°) (a) Calculate the derivatives of these functions.

(b) Deduce that
$$f_1 = f_2 = f_3 = f_4 = f_5.$$

We call this common function f.

(c) Show that f is a bijection from I onto $\mathbb{R}$.

2°) We denote by g the inverse of f.

(a) Give various analytic expressions for $g(x)$.

(b) What is the derivative of g? Compare it with the derivative of f.

---

42. ★ *Uniformly continuous functions do not grow too fast*

1°) Is the function $x \mapsto f(x) = \sin(x^2)$ uniformly continuous on $\mathbb{R}$?

2°) (a) Let $f : \mathbb{R} \to \mathbb{R}$ be uniformly continuous on $\mathbb{R}$. Show that there exist nonnegative real numbers a and b such that:

$$|f(x)| \leqslant a\,|x| + b \text{ for all } x \in \mathbb{R}. \tag{1}$$

(b) Is a continuous function satisfying (1) uniformly continuous on $\mathbb{R}$?

---

43. ★ *A differentiable but not continuously differentiable function.*

Let $f : \mathbb{R} \to \mathbb{R}$ be defined as follows:

$$f(x) = \begin{cases} x^2 \sin\left(\frac{1}{x}\right) & \text{if } x \neq 0, \\ 0 & \text{if } x = 0. \end{cases}$$

Then f is differentiable at any point of $\mathbb{R}$ but its derivative function is not continuous at 0.

---

44. ★ *Slopes with straddling sequences*

Let $f : \mathbb{R} \to \mathbb{R}$ be differentiable at a. Consider two sequences (x_n) and (y_n) converging towards a, but with $x_n < a < y_n$ for all n.

Prove that

$$\frac{f(y_n) - f(x_n)}{y_n - x_n} \to f'(a) \text{ as } n \to +\infty. \tag{1}$$

What happens if one just supposes that the sequences (x_n) and (y_n), with $x_n \neq y_n$ for all n, converge towards a?

---

45. ★ Let $f : \mathbb{R} \to \mathbb{R}$ be a twice differentiable function. We suppose that there exists a $c \in \mathbb{R}$ such that

$$f'(c) \neq \frac{f(b) - f(a)}{b - a} \text{ for all } a \neq b.$$

This may happen. Take for example $f(x) = x^3$ and $c = 0$.

Question: What can then be said about $f''(c)$?

---

46. ★ Let $f : \mathbb{R} \to \mathbb{R}$ be a twice differentiable function. Suppose that $f'(x) \neq 0$ in a neighborhood of a.

Prove that

$$\frac{1}{f(a + h) - f(a)} - \frac{1}{h f'(a)}$$

has a limit when $h(\neq 0) \to 0$.

---

47. ★ Let $f : \mathbb{R} \to \mathbb{R}$ be a thrice continuously differentiable function. Show that there exists a $c \in \mathbb{R}$ for which

$$f(c) \times f'(c) \times f''(c) \times f'''(c) \geqslant 0.$$

---

48. ★ *Infinitely many local minimizers but no local maximizer*

Let $f : \mathbb{R}^2 \to \mathbb{R}$ be defined as

$$f(x, y) = x^2 - \sin y.$$

Check that there are infinitely many local minimizers but no local maximizer.

---

49. ★ *A "pocket proof" of the* LEBESGUE–RIEMANN *lemma*

Let $f : [0, \pi] \to \mathbb{R}$ be a continuous function. We intend to prove that

$$\int_0^\pi f(x)\sin(nx)\,dx \;\to\; 0 \text{ as } n \to +\infty; \tag{1-1}$$

$$\int_0^\pi f(x)\cos(nx)\,dx \;\to\; 0 \text{ as } n \to +\infty. \tag{1-2}$$

For that, let us consider

$$b_n = \frac{2}{\pi}\int_0^\pi f(x)\sin(nx)\,dx,$$

$$S_n = \int_0^\pi \left[f(x) - \sum_{k=1}^n b_k \sin(kx)\right]^2 dx,$$

where n is a positive integer.

1°) Prove that

$$S_n = \int_0^\pi f^2(x)\,dx - \frac{2}{\pi}\sum_{k=1}^n b_k^2.$$

2°) By observing that the sequence (S_n) is decreasing and bounded from below, obtain the result (1-1).

---

50. ★ *Equalities after increasing by small steps*

Let $f : [0, 1] \to \mathbb{R}$ be a continuous function satisfying $f(0) = f(1)$. Let n be a positive integer.

Show that there exists a c in $[0, 1 - \frac{1}{n}]$ such that

$$f\left(c + \frac{1}{n}\right) = f(c).$$

---

51. ★ ROLLE's *theorem on an unbounded interval*

Let $f : [a, +\infty) \to \mathbb{R}$ be continuous on $[a, +\infty)$ and differentiable on $(a, +\infty)$. We suppose that $f(a) = A$ and that $f(x) \to A$ when $x \to +\infty$.

Show that there exists a $c \in (a, +\infty)$ such that $f'(c) = 0$.

---

52. ★ Let p and q be two real numbers and let n be a positive integer. Show that the polynomial function

$$P_n(x) = x^n + px + q$$

has at most two real roots if n is even, and at most three real roots if n is odd.

---

53. ★ *When the mean value theorem "almost" applies*

1°) Let $f : [0, 1] \to \mathbb{R}$ be continuous on $[0, 1]$ and differentiable on $(0, 1)$. Prove that it is impossible to have:

$$\begin{cases} f(0) = 0, \ f(1) = 0; \\ |f'(x)| = 1 \text{ for all } x \in (0, 1). \end{cases} \tag{1}$$

2°) We modify the second part of the conditions in (1) by assuming that f is differentiable for all $x \in (0, 1)$ except at one point, and that $|f'(x)| = 1$ for all x where f is differentiable. Is such a situation possible?

---

54. ★ Let $f : [a, b] \to (0, +\infty)$ be continuous on $[a, b]$ and differentiable on (a, b). Show that there exists a $c \in (a, b)$ such that

$$\frac{f(b)}{f(a)} = \exp\left[\frac{f'(c)}{f(c)}(b - a)\right].$$

---

55. ★ Let $f : [a, b] \to \mathbb{R}$ be continuous on $[a, b]$ and differentiable on (a, b). We suppose moreover that $f(a) = f(b) = 0$.

Show that, for any $k \in \mathbb{R}$, there exists a $c \in (a, b)$ such that

$$f'(c) = kf(c).$$

---

56. ★ CAUCHY's *mean value theorem*

Let $f, g : [a, b] \to \mathbb{R}$ be continuous on $[a, b]$ and differentiable on (a, b). Show that there exists a $c \in (a, b)$ such that

$$\det\begin{bmatrix} f(b) - f(a) & f'(c) \\ g(b) - g(a) & g'(c) \end{bmatrix} = 0.$$

---

57. ★ *Convex functions are (almost) never bounded from above*

Let $f : \mathbb{R}^n \to \mathbb{R}$ be a convex function. Suppose that f is bounded from above on $\mathbb{R}^n$. Show that f is necessarily a constant function.

---

58. ★ *Minimizing a sum of distances to a collection of real numbers*

The following minimization problem appears in Statistics or Optimization. Let $a_1, a_2, ..., a_n$ be distinct real numbers.

Find those x in $\mathbb{R}$ minimizing

$$f(x) = \sum_{i=1}^{n} |x - a_i|$$

on the whole of $\mathbb{R}$.

---

59. ★ *An n-th derivative (1)*

Let $f : (0, +\infty) \to \mathbb{R}$ be defined by: $f(x) = \frac{1}{x}$.

Show that the n-th derivative of f is

$$f^{(n)}(x) = \frac{(-1)^n n!}{x^{n+1}}.$$

---

60. ★ *An n-th derivative (2)*

Let $f : (0, +\infty) \to \mathbb{R}$ be defined by: $f(x) = x^n \log(x)$, where n is a given positive integer.

1°) Prove, by induction for example, that the k-th derivative of f, for a positive integer k, is as follows:

$$f^{(k)}(x) = x^{n-k} n(n-1)...(n-k+1) \left[\frac{1}{n} + \frac{1}{n-1} + ... \frac{1}{n-k+1} + \log(x) \right].$$

2°) Deduce from this $f^{(n)}(\frac{1}{n})$, as well as the limit value of $\frac{1}{n!} f^{(n)}(\frac{1}{n})$ as $n \to +\infty$.

---

61. ★ *A uniqueness result for first-order linear differential equations*

We suppose here that we do not know anything about the exponential function.

1°) Let f be a differentiable function from $\mathbb{R}$ into $\mathbb{R}$ satisfying $f' = f$. Show that the function h_1 from $\mathbb{R}$ into $\mathbb{R}$ defined by $h_1(x) = f(x)f(-x)$ is constant on $\mathbb{R}$.

14

2°) Let f and g be two differentiable functions from $\mathbb{R}$ into $\mathbb{R}$ satisfying $f' = f$ and $g' = g$. Show that the function h_2 from $\mathbb{R}$ into $\mathbb{R}$ defined by $h_2(x) = f(x)g(-x)$ is constant on $\mathbb{R}$.

3°) Deduce from the results above that the solution of the following CAUCHY problem

$$(\mathcal{C}) \quad \begin{cases} y' = y, \\ y(0) = 1 \end{cases}$$

is unique.

---

62. ★ *A uniqueness result for second-order linear differential equations*

Consider the following differential equation, a CAUCHY problem in fact:

$$(\mathcal{C}) \quad \begin{cases} y'' + \omega^2 y = 0, \\ y(0) = 0, \\ y'(0) = 0. \end{cases}$$

We intend to show that only the null function is a solution of $(\mathcal{C})$. For that purpose, let $\varphi : \mathbb{R} \to \mathbb{R}$ be defined as follows:

$$\varphi(t) = [y'(t)]^2 + \omega^2 [y(t)]^2 .$$

Verify that φ' is the zero function, and that $\varphi(0) = 0$. Consequently, φ is the null function.

---

63. ★ *Periodic solutions of a first-order linear differential equation*

Let $a : \mathbb{R} \to \mathbb{R}$ be a continuous function, assumed periodic of period $T > 0$. We consider the following CAUCHY problem:

$$(\mathcal{C}) \quad \begin{cases} x' = a(t)\, x, \\ x(0) = x_0, \end{cases}$$

where $x_0 \neq 0$.

1°) Show that the solution of $(\mathcal{C})$ is periodic of period T if and only if $\int_0^T a(s)\, ds = 0$.

2°) Illustrate this with:

$$(\mathcal{C}_2) \quad \begin{cases} x' = \sin(t)\, x, \\ x(0) = x_0, \end{cases}$$

$$(\mathcal{C}_3) \quad \begin{cases} x' = [1 + \cos(t)]\, x, \\ x(0) = x_0. \end{cases}$$

Is the solution of $(\mathcal{C}_2)$ 2π-periodic? Is the solution of $(\mathcal{C}_3)$ 2π-periodic?

64. ★ *A simple commutative (or Abelian) group*

We define on $I = (-1, 1)$ the following binary operation:

$$(x \in I, \ y \in I) \mapsto x * y = \frac{x + y}{1 + xy}.$$

After having checked that $x * y$ indeed belongs to I, show that $(I, *)$ is a commutative group.

---

65. ★ *A limit of mean values of lengths of chords in a regular polygon*

Let $A_0, A_1, \ldots , A_{n-1}$ be the n vertices of a regular polygon inscribed in the unit circle in the plane, with $A_0 = (1, 0)$ for example. We consider the mean value of the lengths $A_0 A_1, \ldots , A_0 A_{n-1}$:

$$L_n = \frac{1}{n-1}(A_0 A_1 + A_0 A_2 + \ldots + A_0 A_{n-1}).$$

Calculate the limit of L_n as $n \to +\infty$.

---

66. ★ *Expressing a product of lengths of chords in a regular polygon*

Let $S_1, S_2, \ldots , S_n$ be the n vertices of a regular polygon inscribed in the unit circle in the plane. We connect S_1 to S_2, S_1 to $S_3, \ldots ,$ S_1 to S_n, which gives rise to $n - 1$ line-segments of lengths $l_2, l_3, \ldots , l_n$.

Question: express $L = \prod\limits_{k=2}^{n} l_k$ as simply as possible as a function of n.

---

67. ★ *Solving a nonlinear system of 3 equations with 3 variables*

Find the positive real numbers x, y, z satisfying:

$$\begin{cases} x + y + z = 1, \\ \frac{1}{x} + \frac{1}{y} + \frac{1}{z} = 1, \\ x^2 + y^2 + z^2 = 9. \end{cases}$$

---

68. ★ *Linear mappings: mixing dimensions of kernels and those of images*

Let $A : \mathbb{R}^n \to \mathbb{R}^n$ be a linear mapping.

Show that, for every positive integer p, we have:

$$\dim(\ker A^{p+1}) = \dim(\ker A) + \sum_{k=1}^{p} \dim(\operatorname{Im} A^p \cap \ker A);$$

$$\dim(\operatorname{Im} A^{p+1}) = \dim(\operatorname{Im} A) - \sum_{k=1}^{p} \dim(\operatorname{Im} A^p \cap \ker A).$$

---

69. ★ *On the definition of invertibility of a matrix*

Let $A \in \mathcal{M}_n(\mathbb{R})$. Two students are talking about the invertibility property of A.
One says that A is invertible if and only if there exists a $B \in \mathcal{M}_n(\mathbb{R})$ for which

$$AB = BA = I_n. \tag{1}$$

The other one claims that it is enough to have $B \in \mathcal{M}_n(\mathbb{R})$ such that

$$AB = I_n. \tag{2}$$

They are arguing about this, in particular, because multiplication of matrices is not commutative... Who is right?

---

70. ★ *Expected value of the determinant of a random matrix*

Let $M = [m_{i,j}] \in \mathcal{M}_n(\mathbb{R})$ be a matrix whose entries take randomly the values 1 or -1; more precisely, the n^2 entries $m_{i,j}$ are independent random variables obeying the same law:
$$P(m_{i,j} = 1) = P(m_{i,j} = -1) = \frac{1}{2}.$$

Therefore, the determinant of M is a new integer-valued random variable. Question: what is the expectation of $\det M$?

---

71. ★ PAULI *matrices in Electronics*

The treatment of spins of particles in Electronics leads to the so-called PAULI matrices; here they are:
$$\begin{bmatrix} 0 & 1 \\ 1 & 0 \end{bmatrix}, \begin{bmatrix} 1 & 0 \\ 0 & -1 \end{bmatrix}, \begin{bmatrix} 0 & -i \\ i & 0 \end{bmatrix}.$$

The first two matrices are real symmetric and the third is complex Hermitian. They all have the same eigenvalues: -1 and 1. We now consider more specifically the matrix $A = \begin{bmatrix} 0 & -i \\ i & 0 \end{bmatrix}$.

1°) After checking that A is unitary, determine A^{-1} and A^p for positive integers p.

2°) Propose a unitary matrix U satisfying

$$U^{-1}AU = U^*AU = \begin{bmatrix} -1 & 0 \\ 0 & 1 \end{bmatrix}.$$

---

72. ★ *The* CAUCHY–SCHWARZ *inequality in brief*

Let $\mathbb{R}^n$ be equipped with the usual inner-product and the associated Euclidean norm. Let x and y be two non-null vectors in $\mathbb{R}^n$. We posit

$$A(x,y) = \frac{1}{\|y\|^2} \left\| \|y\|^2 x - \left(x^T y\right) y \right\|^2. \tag{1}$$

1°) Show that

$$A(x,y) = \|x\|^2 \|y\|^2 - \left(x^T y\right)^2. \tag{2}$$

2°) (*a*) Deduce from the above the classical CAUCHY–SCHWARZ inequality:

$$\left|x^T y\right| \leqslant \|x\| \times \|y\|. \tag{3}$$

(*b*) Prove that if $A(x,y) = 0$, then $\frac{x}{\|x\|} = \pm\frac{y}{\|y\|}$. What does this mean for the vectors x and y?

---

73. ★ *Diagonally dominant symmetric matrices*

Let $A \in \mathcal{M}_n(\mathbb{R})$ be a symmetric matrix satisfying the relation

$$a_{i,i} \geqslant \sum_{j=1}^{n} |a_{i,j}| \text{ for all } i = 1, ..., n.$$

Prove that A is positive definite.

---

74. ★ *Functions comparable around a point have the same gradient at this point*

Let $f, g : \mathbb{R}^n \to \mathbb{R}$ be assumed differentiable at a point $a \in \mathbb{R}^n$, coinciding at this point (*i.e.*, $f(a) = g(a)$), and comparable in a neighborhood of a:

$$f(x) \geqslant g(x) \text{ for all } x \text{ in a neighborhood of } a$$

or

$$f(x) \leqslant g(x) \text{ for all } x \text{ in a neighborhood of } a.$$

Show that f and g have the same gradient (vector) at a: $\nabla f(a) = \nabla g(a)$.

---

75. ★ *When are a, b, c the lengths of the sides of a triangle?*

Let $a > 0$, $b > 0$, $c > 0$ be given. We wish to find necessary and sufficient conditions on a, b, c such that the triangle ABC with $BC = a, AC = b, AB = c$ is non-degenerate.

Show that the following three statements are equivalent:

(*i*) a, b, c are the lengths of the sides of a non-degenerate triangle.

(*ii*) $a < b + c$; $b < c + a$; $c < a + b$ (the so-called triangle inequalities).

(*iii*) $|b - c| < a < b + c$.

"If you cannot solve a problem, then there is an easier problem you can solve; find it"

G. POLYA (1887–1985)

76. ★★ *The* KY FAN *inequality*

Consider real numbers $x_1, x_2, ..., x_n$ all lying in the interval $(0, \frac{1}{2}]$. Then, the "complementary to 1" real numbers $y_i = 1 - x_i$ all lie in the interval $[\frac{1}{2}, 1)$.

Prove the following inequality:

$$\frac{\prod_{i=1}^{n} x_i}{\left(\sum_{i=1}^{n} x_i\right)^n} \leqslant \frac{\prod_{i=1}^{n} y_i}{\left(\sum_{i=1}^{n} y_i\right)^n}. \tag{1}$$

---

77. ★★ *Estimation of harmonic numbers*

For an integer $n \geqslant 1$, let $H_n = \sum_{k=1}^{n} \frac{1}{k}$ denote the so-called harmonic (real) numbers.

1°) Show that

$$\ln(n) + \frac{1}{n} < H_n < \ln(n) + 1. \tag{1}$$

2°) Deduce from (1) that $\frac{H_n}{\ln(n)}$ tends to 1 as n goes to $+\infty$.

---

78. ★★ *Convergence of "packages" of harmonic numbers*

For positive x, the mean value inequality applied to the logarithm function implies that

$$\frac{1}{x+1} < \ln(x+1) - \ln(x) < \frac{1}{x}. \tag{1}$$

Use this inequality to find the limit, as n tends to $+\infty$, of the following "summation blocks" or "packages" of harmonic numbers

$$S_p(n) = \sum_{k=n+1}^{pn} \frac{1}{k},$$

where $p \geqslant 2$ is an integer.

---

79. ★★ *Behavior of the "waiting integer" in the summation of the harmonic series*

For integers $k \geqslant 1$, consider the harmonic numbers

$$H_k = \sum_{i=1}^{k} \frac{1}{i}. \tag{1}$$

We know that $H_k \to +\infty$ as $k \to +\infty$.

For a positive integer n, we denote by k_n the first integer k for which $H_k \geqslant n$ (it is thus called the "waiting integer" in the summation in (1)).

1°) (a) What are k_1, k_2, k_3?

(b) What is the behavior of k_n as $n \to +\infty$?

2°) Show that $\frac{k_{n+1}}{k_n}$ has a limit as $n \to +\infty$.

---

80. ★★ *Harmonic numbers are not integers*

For integers $n \geqslant 1$, let $H_n = \sum_{k=1}^{n} \frac{1}{k}$ denote the so-called harmonic numbers.

1°) Show that

$$H_{2^n} \geqslant 1 + \frac{n}{2} \text{ for all } n \geqslant 1.$$

2°) Prove that, except for $n = 1$, the rational number H_n is never an integer.

---

81. ★★ *Limit of differences at infinity vs limits of slopes at infinity*

Let $f : (0, +\infty) \to \mathbb{R}$ be a continuous function. We suppose that the difference $f(x+1) - f(x)$ has a limit l as $x \to +\infty$.

Prove that the slope $\frac{f(x)}{x}$ has the same limit l when $x \to +\infty$.

Illustrate the result with the functions $\ln(x)$, $\sqrt{x}$, $\arctan(x)$, etc.

---

82. ★★ *When products and sums of three integers are equal*

Determine the positive integers x, y, z for which

$$xyz = x + y + z?$$

---

83. ★★ *Estimation of factorial numbers*

1°) Using the "physical" meaning of $\int_1^n \ln(x)\,dx$ (*i.e.*, the area under the graph of the logarithm function and the Ox axis, limited by the vertical axes $x = 1$ and $x = n$), show that

$$\ln\left((n-1)!\right) < \int_1^n \ln(x)dx < \ln(n!).$$

2°) Deduce from the above and the calculation of $\int_1^n \ln(x)dx$ that:

$$e\left(\frac{n}{e}\right)^n < n! < ne\left(\frac{n}{e}\right)^n. \tag{1}$$

---

84. ★★ *Expressing a maximum of real numbers from successive partial minima*

Let $x_1, x_2, \dots, x_n$ be real numbers. Show that

$$\max(x_1, x_2, \dots, x_n) = \sum_{k=1}^n (-1)^{k+1} \sum_{1 \leqslant i_1 < \dots < i_k \leqslant n} \min(x_{i_1}, x_{i_2}, \dots, x_{i_k}). \tag{1}$$

---

85. ★★ *Determining the number of positive integer solutions satisfying a linear equation or a linear inequality*

Let n be a positive integer. Let us consider an integer k satisfying $1 \leqslant k \leqslant n$.

1°) Determine the number of solutions of the linear equation

$$x_1 + x_2, \dots + x_k = n$$

which are positive integers $x_1, x_2, \dots, x_k$.

2°) Determine the number of solutions of the linear inequality

$$x_1 + x_2, \dots + x_k \leqslant n$$

which are positive integers $x_1, x_2, \dots, x_k$.

---

86. ★★ *Inequalities involving square roots of integers*

1°) Let $n \geqslant 5$. Show that

$$\sum_{k=1}^n \frac{1}{\sqrt{k}} > \sqrt{2n}. \tag{1}$$

2°) (*a*) Show that for all positive integers n

$$\frac{1}{2\sqrt{n+1}} < \sqrt{n+1} - \sqrt{n} < \frac{1}{2\sqrt{n}}. \tag{2}$$

21

(b) Application 1. Show that for all positive integers n

$$\sum_{k=1}^{n} \frac{1}{\sqrt{k}} > 2\left(\sqrt{n+1} - 1\right).$$

(c) Application 2. Show that the sequence $(u_n)_{n \geqslant 1}$ defined by

$$u_n = \sum_{k=1}^{n} \frac{1}{\sqrt{k}} - 2\sqrt{n}$$

is strictly decreasing and bounded from below, hence convergent.
Deduce from the above:

$$\sum_{k=1}^{n} \frac{1}{\sqrt{k}} \backsim 2\sqrt{n} \text{ as } n \to +\infty.$$

---

87. ★★ *Expressing $x \in [-1, 1]$ as an infinite sum of terms $1/2^n$*

Let S be the collection of sequences $(\varepsilon_n)_{n \geqslant 1}$ where $\varepsilon_n = \pm 1$. With $(\varepsilon_n)_{n \geqslant 1}$ in S we associate the sequence $(u_n)_{n \geqslant 1}$ defined as follows:

$$u_n = \frac{\varepsilon_1}{2} + \frac{\varepsilon_1 \varepsilon_2}{2^2} + \ldots + \frac{\varepsilon_1 \varepsilon_2 \ldots \varepsilon_n}{2^n}.$$

1°) Show that the sequence $(u_n)_{n \geqslant 1}$ is convergent and that its limit lies in $[-1, 1]$.

2°) Given $x \in [-1, 1]$, prove that there exists an $(\varepsilon_n)_{n \geqslant 1}$ in S such that the associated sequence $(u_n)_{n \geqslant 1}$ converges to x. In short,

$$x = \sum_{n=1}^{+\infty} \frac{\varepsilon_1 \varepsilon_2 \ldots \varepsilon_n}{2^n}.$$

---

88. ★★ *The simplest way to show that* $\sum_{n=1}^{+\infty} \frac{1}{n^2} < +\infty.$

Let, for $n \geqslant 1$, $u_n = \frac{1}{n^2}$.

1°) Show that, whenever $n \geqslant 2$,

$$u_n \leqslant \frac{1}{n(n-1)} = \frac{1}{n-1} - \frac{1}{n}.$$

2°) Deduce from the above that, for all $N \geqslant 1$,

$$\sum_{n=1}^{N} u_n \leqslant 2.$$

---

89. ★★ *Sums of factorials vs the factorial of the last term*

We wish to compare the behavior, as n goes to $+\infty$, of $u_n = \sum_{k=1}^{n} k!$ and $v_n = n!$.

Show that the two sequences are actually equivalent, that is to say:

$$\lim_{n \to +\infty} \frac{u_n}{v_n} = 1.$$

---

90. ★★ Let $(u_n)_n$ be a sequence of real numbers defined recursively as follows:

$$u_0 = 0, \ u_1 = 2;$$
$$u_{n+2} = 2(2n+3)^2 u_{n+1} - 4(n+1)^2(2n+1)(2n+3)u_n \text{ for all } n.$$

Show that u_n can be expressed in a closed form as a function of n.

---

91. ★★ Let $(u_n)_n$ be a sequence of real numbers such that the sequence $(v_n = u_{n+1} - u_n)_n$ converges to l.

Prove that, as $n \to +\infty$,

$$\frac{u_n}{n} \to l \text{ and } \frac{u_1 + u_2 + ... + u_n}{n^2} \to \frac{l}{2}.$$

---

92. ★★ *"Almost"* CAUCHY *sequences*

Let $(u_n)_{n \geqslant 1}$ be a sequence of real numbers satisfying the following property:

$$\text{For all positive integers } p, \ \lim_{n \to +\infty} (u_{n+p} - u_n) = 0. \qquad (1)$$

Is such a sequence convergent?

---

93. ★★ *Subadditive sequences*

A sequence $(u_n)_{n \geqslant 1}$ of nonnegative real numbers is called subadditive if

$$u_{m+n} \leqslant u_m + u_n \text{ for all positive integers } m \text{ and } n.$$

1°) Is such a sequence necessarily convergent?
2°) Show that the sequence $\left(\frac{u_n}{n}\right)_{n \geqslant 1}$ is convergent.

94. ★★ Let (u_n) be a sequence of real numbers. Prove the following equivalence:

$$\left(\lim_{n \to +\infty} u_n = l \right) \iff \left(\lim_{n \to +\infty} (2u_{n+1} - u_n) = l \right). \tag{1}$$

---

95. ★★ Let us consider the series with general term

$$u_n = \frac{1}{1^2 + 2^2 + \dots n^2}, \, n \geqslant 1.$$

Check that this series is convergent and calculate its sum $\sum_{n=1}^{+\infty} u_n$.

---

96. ★★ *Summing the alternating harmonic series*

1°) Let n be a positive integer. Show that

$$\sum_{k=1}^{n} \frac{1}{n+k} = \sum_{k=1}^{2n} \frac{(-1)^{k+1}}{k}. \tag{1}$$

This is called the E. CATALAN identity.

2°) Assuming that the series with general term $(-1)^{k+1}/k$ is convergent, determine the sum $S = \sum_{k=1}^{+\infty} \frac{(-1)^{k+1}}{k}$.

---

97. ★★ *Lemma of* B. MARTINET

In studying the convergence of an iterative process for approaching fixed points of a mapping, we may be faced with three bounded sequences of nonnegative real numbers $(a_n), (b_n), (\varepsilon_n)$ satisfying the following:

$$\lim_{n \to +\infty} \varepsilon_n = 0; \tag{1}$$
$$a_{n+1} \leqslant a_n - b_n + \varepsilon_n \text{ for all } n. \tag{2}$$

Typically, in Approximation theory, a_n is the distance of an approximated fixed point to the set of fixed points (of some mapping), b_n is a term serving to control the convergence, and ε_n is an error term.

1°) Suppose that the series with general term ε_n is convergent.

Show that the series with general term b_n is convergent and that the sequence (a_n) is convergent.

2°) Suppose that the series with general term ε_n is divergent.

Show that there exists a subsequence of (b_n) converging to 0.

---

98. ★★ In studying the convergence of iterative processes in Approximation or Optimization, we may be faced with the following two situations.

1°) First situation. (b_n) is a sequence of nonnegative real numbers, and (α_n) and (β_n) are sequences of real numbers, satisfying for all n:

$$b_{n+1} \leqslant (1 - \alpha_n)b_n + \beta_n;$$
$$\alpha_n \in (0, 1];$$
$$\sum_{n=1}^{+\infty} \alpha_n = +\infty, \quad \lim_{n \to +\infty} \frac{\beta_n}{\alpha_n} = 0.$$

2°) Second situation. (b_n) is a sequence of nonnegative real numbers, and (α_n) and (β_n) are sequences of real numbers, satisfying for all n:

$$b_{n+1} \leqslant (1 - \alpha_n)b_n + \beta_n;$$
$$\alpha_n \geqslant 0;$$
$$\sum_{n=1}^{+\infty} \alpha_n = +\infty, \quad \sum_{n=1}^{+\infty} \beta_n < +\infty.$$

Prove that, in both cases, the sequence (b_n) converges to 0.

---

99. ★★ *Iterating the average of a complex number and its modulus*

Let $(z_n)_{n \geqslant 0}$ be the sequence of complex numbers defined as follows:

$$\text{initial point } z_0 = re^{i\theta};$$
$$\text{for all } n, \ z_{n+1} = \frac{1}{2}(z_n + |z_n|).$$

1°) Express z_n in a closed form as a function of r, θ, n.

2°) Prove that the sequence $(z_n)_n$ is convergent and determine its limit in terms of r and θ.

---

100. ★★ *On the set of limit points of a non-converging sequence*

Let (x_n) be a bounded sequence of real numbers. Its smallest limit point is called a, its largest one b. We assume that (x_n) is not convergent. There thus exists a subsequence of (x_n) which converges to a and another subsequence of (x_n) which converges to b. But what about the other points lying in the line-segment $[a, b]$?

We suppose that $(x_{n+1} - x_n)$ converges to 0 as $n \to +\infty$.

Prove that any intermediate point c in $[a, b]$ is a limit point of (x_n).

---

101. ★★ Let (x_n) be a sequence of real numbers such that:

$$(x_{n+1} - x_n) \text{ converges to } 0 \text{ as } n \to +\infty;$$
$$x_n \to +\infty \text{ as } n \to +\infty.$$

Show that there exists a subsequence $(x_{n_k})_k$ of (x_n) such that

$$x_{n_k} - k \text{ converges to } 0 \text{ as } k \to +\infty.$$

---

102. ★★ *Nonlinearly transferring rationals into rationals and irrationals into irrationals*

Does there exist a nonlinear continuously differentiable function $f : \mathbb{R} \to \mathbb{R}$ such that:

- for any rational x, $f(x)$ is also rational,   (1)
- for any irrational x, $f(x)$ is also irrational?   (2)

---

103. ★★ *The so-called "lemma of prisoners"*

Let (z_n) be a bounded sequence of complex numbers. We suppose that the cluster points of (z_n) belong to a finite set $\{l_1, l_2, \dots, l_p\} \subset \mathbb{C}$ (we assume that $l_i \neq l_j$ for $i \neq j$). Let $\delta > 0$ be chosen satisfying

$$\delta < \min_{i \neq j} |l_i - l_j|.$$

We suppose that $|z_{n+1} - z_n| \leqslant \delta$ for n large enough.

Prove that the sequence (z_n) is convergent and that its limit is one of the $l_i's$.

---

104. ★★ *Anti-Lipschitzian functions are very special*

1°) A function $f : [0,1] \to [0,1]$ is said to be anti-Lipschitzian (or a dilation) if:

$$|f(x) - f(y)| \geqslant |x - y| \text{ for all } x, y \text{ in } [0,1]. \qquad (1)$$

In contrast to the LIPSCHITZ property on $[0,1]$, this requirement on f is stringent.

Determine all the functions which are anti-Lipschitzian on $[0,1]$.

2°) Let $f : \mathbb{R} \to \mathbb{R}$ be continuous and anti-Lipschitzian on $\mathbb{R}$ (*i.e.*, satisfying $|f(x) - f(y)| \geqslant |x - y|$ for all x, y in $\mathbb{R}$). Show that the range of f is the whole of $\mathbb{R}$.

105. ★★ *Oddness and evenness of shifted versions imply periodicity*

We note that the sine function is odd while its shifted version $x \mapsto \sin\left(x + \frac{\pi}{2}\right) = \cos(x)$ is even; both are 2π-periodic. We actually have a general result of this kind.

For a function $f : \mathbb{R} \to \mathbb{R}$, consider its shifted versions $f_r : x \mapsto f_r(x) = f(x + r)$, where r is a real number.

Suppose that there exists two different real numbers r and s, say with $r < s$, such that f_r is even and f_s is odd. Show that f is necessarily periodic. Provide a period $T > 0$ of f in terms of the numbers r and s.

---

106. ★★ *A classical functional equation: functions transforming sums into products*

We look for differentiable functions $f : \mathbb{R} \to \mathbb{R}$ satisfying:

$$f(x + y) = f(x) \times f(y) \text{ for all } x, y \text{ in } \mathbb{R}. \tag{1}$$

1°) Check that, for such f, one necessarily has either $f(0) = 0$ or $f(0) = 1$.

2°) Check that the only f satisfying (1) and $f(0) = 0$ is the null function.

3°) We suppose here that $f(0) = 1$.

(*a*) Prove, by just using the definition of the derivative of f at 1, that any function satisfying (1) is a solution of a linear first-order differential equation.

(*b*) Deduce from that all the solutions of the functional equation (1).

---

107. ★★ *A double functional equation: functions transforming sums into sums and products into products*

We look for continuous functions $f : \mathbb{R} \to \mathbb{R}$ satisfying:

$$
\begin{aligned}
f(x + y) &= f(x) + f(y) \\
&\text{and} \\
f(xy) &= f(x)f(y) \text{ for all } x, y \text{ in } \mathbb{R}.
\end{aligned}
$$

What are such functions?

---

108. ★★ *Periodic solutions of a functional equation*

We look for functions $f : \mathbb{R} \to \mathbb{R}$ which are 2π-periodic, of class $\mathcal{C}^\infty$, and satisfy the following functional equation:

$$f(2x) = 2\sin(x) \cdot f'(x) \text{ for all } x \in \mathbb{R}. \tag{1}$$

1°) Give simple examples of non-null functions satisfying the above requirements.

27

2°) By analysing the FOURIER (complex) coefficients $c_n(f)$, $n = 0, \pm1, \pm2, ...$, of a possible solution f, prove the following:

$$c_{2n}(f) = 0 \text{ for all } n = \pm1, \pm2, ...$$
$$(2n - 1)c_{2n-1}(f) - (2n + 1)c_{2n+1}(f) = c_n(f) \text{ for all } n = 0, \pm1, \pm2, ...$$
$$c_{2n+1}(f) = c_{-(2n+1)}(f) = 0 \text{ for all } n = 1, 2, ...$$

3°) Deduce the solutions of the posed problem.

---

109. ★★ *A general "product breaking" formula for the cosine function*

We know that

$$\cos a \, \cos b = \frac{1}{2}[\cos(a + b) + \cos(a - b)] \tag{1}$$

$$= \frac{1}{4}[\cos(a + b) + \cos(-a - b) + \cos(a - b) + \cos(-a + b)] \tag{2}$$

(since cosine is an even function).

We intend to generalize these formulas.

Prove that, given real numbers $a_1, a_2, ... , a_n$ ($n \geqslant 2$), we have:

$$\cos a_1 \cos a_2 ... \cos a_n = \frac{1}{2^n} \sum_{\varepsilon_i = +1 \text{ or } -1} \cos(\varepsilon_1 a_1 + \varepsilon_2 a_2 + ... + \varepsilon_n a_n) \tag{3}$$

$$= \frac{1}{2^{n-1}} \sum_{\varepsilon_i = +1 \text{ or } -1} \cos(a_1 + \varepsilon_2 a_2 + ... + \varepsilon_n a_n). \tag{4}$$

---

110. ★★ *A "mixture" of two easy linear first-order differential equations*

1°) First solve the two following easy linear CAUCHY problems:

$$(\mathcal{C}_1) \quad x'(t) = t; \; x(0) = 0.$$
$$(\mathcal{C}_2) \quad x'(t) = x(t); \; x(2) = 2.$$

2°) Now solve the following nonlinear CAUCHY problem, a kind of "mixture" of the two previous ones:

$$(\mathcal{C}_3) \quad x'(t) = \max[t, x(t)]; \; x(0) = 0.$$

---

111. ★★ *Extending a function by continuity*

Let $f : (0, 1) \to \mathbb{R}$ be defined as:

$$f(x) = \int_{x^2}^{x} \frac{dt}{\ln(t)}.$$

1°) Check that f is differentiable on $(0, 1)$ and determine the derivative $f'(x)$ of f at $x \in (0, 1)$.

2°) Deduce that f can be extended by continuity to the boundary values $x = 0$ and $x = 1$.

---

112. ★★ Let $f : I \to \mathbb{R}$ be a real-valued function defined on an open interval I of $\mathbb{R}$. Either prove or disprove (with the help of a counterexample, in the latter case) the following statements:

1°) If for every x in I we have

$$\lim_{h \to 0} [f(x+h) - f(x-h)] = 0, \tag{1}$$

then f is continuous on I.

2°) If for every x in I we have

$$\lim_{h \to 0} \frac{f(x+h) - f(x-h)}{h} = 0, \tag{2}$$

then f is constant on I.

Consider the same question if we assume furthermore that f is continuous on I.

3°) If f is continuous on I and if for every x in I we have

$$\lim_{h \to 0} \frac{f(x+h) - 2f(x) + f(x-h)}{h^2} = 0, \tag{3}$$

then f is an affine function, *i.e.* of the form $x \in I \mapsto f(x) = ax + b$.

---

113. ★★ *An easy way to obtain the value of $\int_0^{+\infty} e^{-x^2} dx$*

Let

$$I_R = \iint\limits_{[0,R] \times [0,R]} e^{-(x^2 + y^2)} \, dx dy,$$

$$J_R = \iint\limits_{\Delta_R} e^{-(x^2 + y^2)} \, dx dy, \text{ where } \Delta_R = \left\{ (x, y) : x \geqslant 0, y \geqslant 0, x^2 + y^2 \leqslant R^2 \right\}.$$

1°) Calculate J_R via a change of variables.

2°) Compare J_R, I_R and $J_{R\sqrt{2}}$.

3°) Deduce from the above that $\int_0^R e^{-x^2} dx$ has a limit as $R \to +\infty$ (the limit denoted by $\int_0^{+\infty} e^{-x^2} dx$), whose value is to be determined.

---

114. ★★ *When the convergence of the integrals of $|f|$ and $|f'|^2$ on $[0, +\infty)$ induce the convergence of f towards 0*

Let $f : [0, +\infty) \to \mathbb{R}$ be a continuously differentiable function such that

$$\int_0^{+\infty} |f(x)| \ dx < +\infty \quad \text{and} \quad \int_0^{+\infty} |f'(x)|^2 \ dx < +\infty.$$

Prove that $f(x)$ tends to 0 as $x \to +\infty$.

---

115. ★★ *Pointwise convergence on $[0, 1]$ of polynomial functions implies uniform convergence*

Let (P_n) be a sequence of polynomial functions of degree at most d (d is fixed). We suppose that

$$\lim_{n \to +\infty} P_n(x) = 0 \text{ for all } x \in [0, 1].$$

Show that the convergence of (P_n) towards 0 is uniform on $[0, 1]$.

---

116. ★★ *Limit of an integral*

Let $f : [0, 1] \to \mathbb{R}$ be a continuous function. We set

$$I_n = (n+1) \int_0^1 x^n f(x) \ dx. \tag{1}$$

Show that I_n has a limit as $n \to +\infty$, and determine this limit.

---

117. ★★ Let $f : [0, 1] \to \mathbb{R}$ be a continuous function satisfying:

$$\int_0^1 x^k f(x) \ dx = 0 \text{ for } k = 0, 1, ..., n-1;$$
$$\int_0^1 x^n f(x) \ dx = 0.$$

Prove that there necessarily exists a point $x \in [0, 1]$ at which the value of f is "large", more precisely

$$|f(x)| \geqslant (n+1)2^n.$$

30

118. ★★ *On roots of unitary polynomials*

Let $P(x) = x^n + a_{n-1}x^{n-1} + ... + a_1x - 1$ be a unitary polynomial of degree n with real coefficients $a_{n-1}, ..., a_1$.

We suppose that every root (possibly complex) r of P lies outside the open unit ball of $\mathbb{C}$, *i.e.*, $|r| \geqslant 1$.

Prove that the real number 1 is necessarily a root of P.

---

119. ★★ *The n-th primitive (or antiderivative) function of a continuous function*

Let $f : \mathbb{R} \to \mathbb{R}$ be a continuous function. We know that the primitive (or antiderivative) function F of f, satisfying $F(a) = 0$, can be expressed as:

$$F(x) = \int_a^x f(t)\, dt. \tag{1}$$

Let n be a positive integer. We intend to provide a similar formula for the n-th primitive (or antiderivative) function G of f, that is, the one such that the n-th derivative $G^{(n)}$ of G is f.

Prove that the n-th primitive function G of f, satisfying

$$G(a) = 0, G'(a) = 0, ..., G^{(n-1)}(a) = 0,$$

can be expressed as:

$$G(x) = \int_a^x \frac{(x-t)^{n-1}}{(n-1)!} f(t)\, dt. \tag{2}$$

---

120. ★★ *A particular sine-cosine equation*

What are the (real) solutions of the equation

$$\sin(\cos x) = \cos(\sin x)?$$

---

121. ★★ *When the theorem of composition of differentiable functions (chain rule) does not apply*

Let $f : \mathbb{R} \to \mathbb{R}$ be defined as follows:

$$f(x) = \arcsin\left(\frac{2x}{1+x^2}\right).$$

This function looks nice, it results from the composition of two "very regular" functions: $x \in \mathbb{R} \mapsto \frac{2x}{1+x^2}$ with $y \in [-1,1] \mapsto \arcsin(y)$.

Question: is f differentiable on $\mathbb{R}$?

31

122. ★★ *The limit of an integral*

Let $f : [0, 1] \to \mathbb{R}$ be a continuous function. We then define:

$$\text{For all } x > 0, \ F(x) = \int_0^1 \frac{x \, f(t)}{x^2 + t^2} \, dt.$$

Prove that $F(x)$ has a limit as $x \to 0$ and determine this limit.

---

123. ★★ *The intermediate value property for derivatives*

Let I be an open interval of $\mathbb{R}$ and let $f : I \to \mathbb{R}$ be a differentiable function. We note $A = \{(x, y) \in I \times I \ ; \ x < y\}$.

1°) Check that A is a convex subset of $\mathbb{R}^2$.

2°) For $(x, y) \in A$, define $g(x, y) = \frac{f(y) - f(x)}{y - x}$.

Prove that

$$g(A) \subset f'(I) \subset \overline{g(A)}. \tag{1}$$

3°) (a) Show that $g(A)$ is an interval of $\mathbb{R}$.

(b) Deduce from the result of (a) and (1) that $f'(I)$ is an interval of $\mathbb{R}$.

---

124. ★★ *The square of a derivative function is not necessarily a derivative function*

Let f and g be two functions defined on $\mathbb{R}$ as follows:

$$f(x) = \begin{cases} \sin\left(\frac{1}{x}\right) & \text{if } x \neq 0, \\ 0 & \text{if } x = 0; \end{cases}$$

$$g(x) = \begin{cases} \cos\left(\frac{1}{x}\right) & \text{if } x \neq 0, \\ 0 & \text{if } x = 0. \end{cases}$$

1°) Show that both f and g are derivative functions, that is to say: there are differentiable functions F and G such that $F' = f$ and $G' = g$.

2°) Show that neither f^2 nor g^2 is a derivative function.

---

125. ★★ *Second-order series vs second-order differentiability*

1°) Having a second-order TAYLOR–YOUNG series does not imply second-order differentiability.

Let $f : \mathbb{R} \to \mathbb{R}$ be defined as follows:

$$f(x) = x^3 \sin\left(\frac{1}{x}\right) \ \text{if } x \neq 0; \ f(0) = 0.$$

(a) Check that this function enjoys a second-order series around 0, that is to say:

$$f(x) = f(0) + ax + bx^2 + x^2\varepsilon(x), \tag{1}$$
$$\text{with } \varepsilon(x) \to 0 \text{ as } x \to 0.$$

(b) Verify that f is differentiable on $\mathbb{R}$ but is not twice differentiable at 0.

2°) Converse property.

Let $f : \mathbb{R} \to \mathbb{R}$ be differentiable and convex (or concave) in a neighborhood $(-a, a)$ of 0. Suppose that f enjoys a second-order series like (1). Prove that f is twice differentiable at 0 (and $2b = f''(0)$).

---

126. ★★ *An unusual characterization of differentiability*

Let $f : \mathbb{R} \to \mathbb{R}$ be a function continuous at 0.

1°) Suppose that the quotient $\frac{f(h)-f(-h)}{2h}$ has a limit l as h tends to 0. Does this imply that f is differentiable at 0 (with $l = f'(0)$)?

2°) Let $c \neq 1$. Suppose that f is differentiable at 0. Prove that the quotient $\frac{f(h)-f(ch)}{(1-c)h}$ tends to $f'(0)$ as h tends to 0.

3°) Let $c \neq \pm 1$. We suppose that

$$\frac{f(h) - f(ch)}{(1 - c)h} \text{ has a limit } l \text{ as } h \text{ tends to 0.} \tag{1}$$

Is f then differentiable at 0?

---

127. ★★ *How to differentiate by integrating*

Let $f : \mathbb{R} \to \mathbb{R}$ be a continuous function. For $h > 0$ we define:

$$D_h f(x) = \frac{3}{2h^3} \int_{-h}^{h} t f(x + t)\, dt. \tag{1}$$

1°) A first example. Calculate $D_h f(x)$ for $f = |\cdot|$ and $x = 0$.

2°) (a) Suppose that f is differentiable at x. Show that $D_h f(x) \to f'(x)$ as $h \to 0$.

(b) Generalization. Suppose just that f has a (finite) right-derivative $f'_+(x)$ and a (finite) left-derivative $f'_-(x)$ at x. Prove that

$$\lim_{h \to 0} D_h f(x) = \frac{1}{2}\left[f'_+(x) + f'_-(x)\right]. \tag{2}$$

33

128. **★★** *The* HERMITE–HADAMARD *inequality for convex functions*

Let $f : I \to \mathbb{R}$ be a convex function on the open interval I and let $a < b$ in I. Show that the mean value $\frac{1}{b-a} \int_a^b f(x) \, dx$ of f on $[a,b]$ can be squeezed between values of f at $a, b, (a+b)/2$ as follows:

$$f\left(\frac{a+b}{2}\right) \leqslant \frac{1}{b-a} \int_a^b f(x) \, dx \leqslant \frac{f(a) + f(b)}{2}. \tag{1}$$

---

129. **★★** *Uniform convergence of polynomial functions*

Let $f : \mathbb{R} \to \mathbb{R}$ be a continuous function. Such a function can be uniformly approximated on $[a,b]$ by a polynomial function: indeed, there exists a sequence of polynomial functions (P_n) which converges towards f uniformly on $[a,b]$ (this is the WEIERSTRASS approximation theorem).

Suppose now that we consider uniform convergence on the whole of $\mathbb{R}$. Question: assuming there exists a sequence of polynomial functions (P_n) converging towards f uniformly on $\mathbb{R}$, what can be said about f?

---

130. **★★** *A "second-order" mean value theorem (1)*

Let $k > 0$ be fixed. Determine all the twice differentiable functions $f : \mathbb{R} \to \mathbb{R}$ satisfying the following "second order" mean value property:

$$f(x) - f(y) = (y - x)f'(x) + \frac{(x-y)^2}{2} f''\left(\frac{x + ky}{1 + k}\right) \quad \text{for all } x \text{ and } y \text{ in } \mathbb{R}. \tag{1}$$

---

131. **★★** *A "second-order" mean value theorem (2)*

Determine all the twice differentiable functions $f : \mathbb{R} \to \mathbb{R}$ satisfying the following "second order" mean value property:

$$f(x) - f(y) = \frac{x^2 - y^2}{2} f''\left(\frac{x + y}{2}\right) \quad \text{for all } x \text{ and } y \text{ in } \mathbb{R}. \tag{1}$$

---

132. **★★** Let $f : \mathbb{R} \to \mathbb{R}$ be of class $\mathcal{C}^\infty$. Consider n distinct real points $a_1 < a_2 < ... < a_n$ and suppose that $f(a_i) = 0$ for all $i = 1, 2, ... n$.

Show that for any x lying in the interval $[a_1, a_n]$, there exists a $\theta \in (a_1, a_n)$ such that

$$f(x) = \frac{(x - a_1)(x - a_2)... \, (x - a_n)}{n!} f^{(n)}(\theta).$$

---

133. ★★ *Locally averaging a convex function*

Let $f : \mathbb{R} \to \mathbb{R}$ be a convex function. For $h > 0$, we define a new function $f_h : \mathbb{R} \to \mathbb{R}$ as follows:

$$f_h(x) = \frac{1}{2h} \int_{x-h}^{x+h} f(t) \, dt, \ x \in \mathbb{R}.$$

Prove that f_h is convex.

---

134. ★★ *An involution operation on convex functions of the real variable*

Let $f : (0, +\infty) \to \mathbb{R}$ be a convex function. Let now $f^\diamond : (0, +\infty) \to \mathbb{R}$ be defined as:

$$f^\diamond(x) = x f\left(\frac{1}{x}\right).$$

Show that $f^\diamond$ is convex.

---

135. ★★ *Concave functions and subadditivity*

Let $f : [0, +\infty) \to [0, +\infty)$ be a concave function.

1°) Give examples of such functions.

2°) Show that they are subadditive, that is to say:

$$f(x+y) \leqslant f(x) + f(y) \text{ for all } x, y \text{ in } [0, +\infty).$$

---

136. ★★ *Uniform limit of convex functions*

Let $f_n : \mathbb{R} \to \mathbb{R}$ be a convex function. We suppose that the sequence (f_n) converges pointwise towards a function $f : \mathbb{R} \to \mathbb{R}$.

1°) Show that f is convex.

2°) (a) Prove that the convergence of (f_n) towards f is uniform on any interval $[a, b]$.

(b) Does the result remain true if we assume the pointwise convergence of (f_n) towards f just on $[a, b]$?

---

137. ★★ *Differentiable vs continuously differentiable convex functions*

Let $f : \mathbb{R}^n \to \mathbb{R}$ be a convex function.

1°) Prove that f is differentiable at x if and only if the partial derivatives $\partial f / \partial x_i$ exist at x, for $i = 1, ..., n$.

2°) Show that if f is differentiable in a neighborhood of x, then ∇f is continuous at x.

Hence, if $\Omega \subset \mathbb{R}^n$ is an open convex set, the following equivalence holds true for the convex f :

$$(f \text{ is differentiable on } \Omega) \Leftrightarrow (f \text{ is continuously differentiable on } \Omega). \qquad (1)$$

---

138. ★★ *A functional equation of* D'ALEMBERT (1769)

The constant functions 0 and 1 obviously satisfy the functional equation

$$f(x)f(y) = \frac{1}{2} \left[f(x+y) + f(x-y) \right] \text{ for all real } x \text{ and } y. \qquad (1)$$

What are the other continuous functions $f : \mathbb{R} \to \mathbb{R}$ satisfying (1)?

---

139. ★★ *A functional inequality*

Let $f : \mathbb{R} \to \mathbb{R}$ be a twice continuously differentiable function satisfying:

$$f(x+y)f(x-y) \leqslant [f(x)]^2 \text{ for all } x \text{ and } y \text{ in } \mathbb{R}. \qquad (1)$$

Show that
$$f(x)f''(x) \leqslant [f'(x)]^2 \text{ for all } x \text{ in } \mathbb{R}. \qquad (2)$$

---

140. ★★ *The n-th derivative of a composite function (1)*

Let $f : \mathbb{R} \to \mathbb{R}$ be n times differentiable at x and $g : \mathbb{R} \to \mathbb{R}$ be defined as $g(x) = f(e^x)$.

Show that one can express the n-th derivative of g at x in terms of successive derivatives of f at x as follows:

$$g^{(n)}(x) = \sum_{k=1}^{n} a_k^{(n)} e^{kx} f^{(k)}(e^x),$$

where $a_k^{(n)}$ are real coefficients, not depending on f, which can be determined recursively.

---

141. ★★ *The n-th derivative of a composite function (2)*

Let $f : \mathbb{R} \to \mathbb{R}$ be n times differentiable at x, let $g : (0, +\infty) \to \mathbb{R}$ be defined as $g(x) = f(\log x)$.

Show that one can express the n-th derivative of g at x in terms of successive derivatives of f at x as below:

$$g^{(n)}(x) = \frac{1}{x^n} \sum_{k=1}^{n} a_k^{(n)} f^{(k)}(\log x),$$

where $a_k^{(n)}$ are real coefficients, not depending on f, which can be determined recursively.

---

142. ★★ *Derivatives of the slope function*

Let $f : \mathbb{R} \to \mathbb{R}$ be a function of class $\mathcal{C}^\infty$. We define the slope function of f at 0 as follows:

$$g(x) = \frac{f(x) - f(0)}{x} \text{ if } x \neq 0;$$
$$g(0) = f'(0).$$

1°) Verify that

$$g(x) = \int_0^1 f'(tx) \, dt \text{ for all } x \in \mathbb{R}.$$

2°) Prove that g is a function of class $\mathcal{C}^\infty$, and show how to calculate the derivatives of g in terms of those of f.

---

143. ★★ *Linear independence of functions vs linear independence of their derivatives*

Let $f_1, f_2, \dots, f_n$ be n linearly independent differentiable functions from $\mathbb{R}$ into $\mathbb{R}$. Consider now their derivative functions $f_1', f_2', \dots, f_n'$.

Prove that at least $n - 1$ among them are linearly independent functions.

---

144. ★★ *Periodicity of areas implies periodicity of functions*

Let $f : \mathbb{R} \to \mathbb{R}$ be a continuous function.

1°) We suppose that there exist $T > 0$ and $A \in \mathbb{R}$ such that

$$\int_x^{x+T} f(t) \, dt = A \text{ for all } x \in \mathbb{R}.$$

Show that f itself is periodic of period T.

2°) We suppose that f is periodic of period T. Consider a primitive function (or antiderivative) F of f. Is F periodic of period T?

---

145. ★★ *Minimizing the integral of the deviation of a function from a constant*

Let $f : [a, b] \to \mathbb{R}$ be a strictly increasing and continuous function. For a real number r, we consider the integral of the deviation of f from r as:

$$I(r) = \int_a^b |f(x) - r| \, dx.$$

Show that $I(r)$ is minimized at one, and only one, point r^*.

---

146. ★★ *Impossible mean-value constraints*

Let $\mu \in [0,1]$. We look for continuous functions $f : [0,1] \to \mathbb{R}$ satisfying the following three mean-value constraints:

$$\int_0^1 f(x)\, dx = 1; \tag{1}$$

$$\int_0^1 x f(x)\, dx = \mu; \tag{2}$$

$$\int_0^1 x^2 f(x)\, dx = \mu^2. \tag{3}$$

What are they?

---

147. ★★ Let $f : \mathbb{R} \to \mathbb{R}$ be a continuously differentiable function. We define $g : \mathbb{R}^2 \to \mathbb{R}$ as follows:
$$g(x,y) = \frac{f(x) - f(y)}{x - y} \text{ if } x \neq y; \quad g(x,x) = f'(x).$$

1°) Is g a continuous function?

2°) We suppose that f is twice continuously differentiable on $\mathbb{R}$. Show that g is differentiable on $\mathbb{R}^2$ and determine the gradient vector $\nabla g(x,x)$ of g at (x,x).

---

148. ★★ *The matrix* $\exp(tA)$ *made explicit for* $A \in \mathcal{M}_2(\mathbb{R})$

Let $A \in \mathcal{M}_2(\mathbb{R})$. We intend to make $\exp(tA)$ explicit whatever A is.

1°) Suppose that A has two distinct real eigenvalues λ_1 and λ_2. Show that

$$\begin{aligned}
\exp(tA) &= e^{\lambda_1 t} I_2 + \frac{e^{\lambda_2 t} - e^{\lambda_1 t}}{\lambda_2 - \lambda_1}(A - \lambda_1 I_2) \\
&= e^{\lambda_2 t} I_2 + \frac{e^{\lambda_1 t} - e^{\lambda_2 t}}{\lambda_1 - \lambda_2}(A - \lambda_2 I_2).
\end{aligned}$$

2°) Suppose that A has a double real eigenvalue λ_0. Show that

$$\exp(tA) = e^{\lambda_0 t} I_2 + t e^{\lambda_0 t}(A - \lambda_0 I_2).$$

3°) Suppose that A has two complex conjugate eigenvalues $\lambda_1 = \alpha + i\beta$ and $\lambda_2 = \alpha - i\beta$, with $\beta \neq 0$. Show that

$$\exp(tA) = e^{\alpha t}\left[\frac{\sin(\beta t)}{\beta} A + \left(\cos(\beta t) - \frac{\alpha}{\beta}\sin(\beta t)\right) I_2\right].$$

38

149. ★★ *A second-order matricial linear differential equation*

Let $A \in \mathcal{M}_n(\mathbb{R})$. Consider the following (vectorial) CAUCHY problem:

$$(\mathcal{C}) \quad \begin{cases} Y''(t) = A\, Y(t), \\ Y(0) = M_0, \\ Y'(0) = M_1, \end{cases}$$

where M_0 and M_1 are given matrices in $\mathcal{M}_n(\mathbb{R})$.

We define

$$C(t) = \sum_{k=0}^{+\infty} \frac{t^{2k}}{(2k)!} A^k \text{ and } S(t) = \sum_{k=0}^{+\infty} \frac{t^{2k+1}}{(2k+1)!} A^k. \tag{1}$$

Prove that the unique solution of $(\mathcal{C})$ is

$$t \in \mathbb{R} \mapsto Y(t) = M_0 C(t) + M_1 S(t). \tag{2}$$

---

150. ★★ *A necessarily commutative group*

Let $(G, *)$ be a group, with e as unit element. We suppose that

$$x * x = e \text{ for all } x \text{ in } G.$$

Show that such a group is necessarily commutative (or Abelian).

---

151. ★★ *Nilpotence in a ring*

Let $(\mathcal{R}, +, \times)$ be a unitary ring. The unit element in $\mathcal{R}$ is denoted by the symbol e. An element $x \in \mathcal{R}$ is said to be nilpotent if there is a positive integer p such that $x^p = 0$.

1°) Show that:
$$(e - x \text{ nilpotent}) \implies (x \text{ invertible}).$$

2°) Assume moreover that $\mathcal{R}$ is commutative. Show that:

$$(x \text{ and } y \text{ both nilpotent}) \implies (x \times y, x + y \text{ and } x - y \text{ are nilpotent}).$$

3°) Show with a counterexample that, if $\mathcal{R}$ is not commutative, the sum of two nilpotent elements is not necessarily nilpotent.

39

152. ★★ *The area of a triangle from the lengths of its sides*

Let a, b, c be the lengths of the sides of a triangle. We intend to calculate the area $\mathcal{A}$ of this triangle in terms of a, b, c only.

Prove that

$$\mathcal{A} = \frac{1}{4}\sqrt{(a+b+c)(-a+b+c)(a-b+c)(a+b-c)}. \qquad (1)$$

153. ★★ *Maximizing the product of distances to the sides of a triangle*

Consider an acute triangle $\mathcal{T} = ABC$. For a point P inside $\mathcal{T}$, let us define

$$f(P) = \text{ product of distances from } P \text{ to the sides of } \mathcal{T}.$$

Determine the points P in the triangle where this function f is maximized, as well as the maximal value of f in terms of the lengths AB, AC, BC and the area $\mathcal{A}$ of the triangle.

154. ★★ *A property of perpendicular radiuses in an ellipse*

1°) Let ABC be a right-angled triangle, the right angle being at A. We denote by AH the altitude drawn from A in the triangle.

What is the relation linking the lengths of AB, AC and AH?

2°) Let $\mathcal{E}$ be the ellipse in the plane whose Cartesian equation is

$$\frac{x^2}{a^2} + \frac{y^2}{b^2} = 1,$$

with $a \geqslant b > 0$. The center of the ellipse is the origin denoted by O.

Consider two points M and N on the ellipse such that the triangle OMN is right-angled, the right angle being at O; in other words, the two radiuses OM and ON of the ellipse are perpendicular.

Prove that the line MN is tangent to a circle centered at O and whose radius is to be determined.

155. ★★ *A tetrahedron with faces of equal areas*

Let us consider a tetrahedron whose four faces have the same area.

Show that two opposite edges have the same length.

156. ★★ *A relation involving the roots of a polynomial of degree 3*

Let a, b, c be the roots, supposed distinct, of a polynomial P of degree 3.

1°) Why are $P'(a), P'(b), P'(c)$ necessarily non-zero?

2°) Prove the following relation:

$$\frac{a}{P'(a)} + \frac{b}{P'(b)} + \frac{c}{P'(c)} = 0.$$

---

157. ★★ *Finite groups of (diagonalizable) matrices*

Let G be a finite group of matrices in $\mathcal{M}_n(\mathbb{C})$. The group operation is the usual multiplication of matrices.

Show that every matrix in G is diagonalizable.

---

158. ★★ *Determining the dimension of a specific vector subspace of matrices*

Let $\mathcal{V}$ be the vector subspace of $\mathcal{M}_n(\mathbb{R})$ generated by the set

$$\{AB - BA \mid A \text{ and } B \text{ in } \mathcal{M}_n(\mathbb{R})\}.$$

Determine the dimension of $\mathcal{V}$.

---

159. ★★ *Distances from the canonical basis to a vector subspace*

Let $\mathbb{R}^n$ be the usual Euclidean space and let $(e_1, e_2, \dots, e_n)$ be its canonical basis.

Let V be a vector subspace of $\mathbb{R}^n$ of dimension p. We denote by $d(e_i, V)$ the (Euclidean) distance from e_i to V.

1°) Show that

$$\sum_{i=1}^{n} [d(e_i, V)]^2 = n - p. \tag{1}$$

2°) Show that

$$\max_{i=1,\dots,n} d(e_i, V) = \sqrt{\frac{n-p}{n}} \tag{2}$$

if and only if all the e_i' s are equidistant from V, that is to say:

$$d(e_1, V) = d(e_2, V) = \dots = d(e_n, V).$$

41

160. ★★ *From invertible matrices to arbitrary matrices*

1°) Show that any matrix $M \in \mathcal{M}_n(\mathbb{C})$ can be written as the sum of two invertible matrices in $\mathcal{M}_n(\mathbb{C})$.

2°) Show that any matrix $M \in \mathcal{M}_n(\mathbb{C})$ is the limit of a sequence of invertible matrices in $\mathcal{M}_n(\mathbb{C})$.

---

161. ★★ *There is an invertible matrix in any hyperplane*

Let $\mathcal{H}$ be an arbitrary hyperplane of $\mathcal{M}_n(\mathbb{R})$, $n \geqslant 2$.

Prove that there necessarily exists an invertible matrix in $\mathcal{H}$ (in other words, the open set of invertible matrices of $\mathcal{M}_n(\mathbb{R})$ meets $\mathcal{H}$).

---

162. ★★ *Bounding the distance of an eigenvalue to the average of the eigenvalues*

Let $A = [a_{i,j}] \in \mathcal{M}_n(\mathbb{C})$. The average of the eigenvalues $\lambda_1, \lambda_2, ..., \lambda_n$ of A is

$$\lambda^* = \frac{\lambda_1 + \lambda_2 + ... + \lambda_n}{n} = \frac{\operatorname{tr}(A)}{n}.$$

Let λ_i be any eigenvalue of A. Prove that

$$|\lambda_i - \lambda^*|^2 \leqslant \frac{n-1}{n} \left| \sum_{i,j} |a_{i,j}|^2 - \frac{|\operatorname{tr}(A)|^2}{n} \right|. \tag{1}$$

---

163. ★★ *Continuity of the characteristic polynomial and of the minimal polynomial*

1°) Consider the mapping which associates with $M \in \mathcal{M}_n(\mathbb{C})$ its characteristic polynomial. Is this mapping continuous?

2°) Consider now the mapping which associates with $M \in \mathcal{M}_n(\mathbb{C})$ its minimal polynomial. Is this mapping continuous?

---

164. ★★ *Characterizations of nilpotent matrices*

Recall that a matrix $M \in \mathcal{M}_n(\mathbb{C})$ is said to be nilpotent if there is a positive integer p such that $M^p = 0$. Such a p can be chosen to be at most n, without loss of generality.

1°) Show that M is nilpotent if and only if

$$\operatorname{tr}(M^k) = 0 \text{ for all } k = 1, 2, ..., n. \tag{1}$$

2°) Prove that M is nilpotent if and only if all the eigenvalues of M are 0.

---

165. ★★ *A determinant involving successive derivatives of a function*

Let $f : \mathbb{R} \to \mathbb{R}$ be a function of class $\mathcal{C}^\infty$. Consider the following $n \times n$ matrix, built up from the successive derivatives of f like in TAYLOR expansions,

$$
M_n = \begin{bmatrix}
f' & f & 0 & 0 & . & . & 0 \\
\frac{f''}{2!} & f' & f & 0 & . & . & 0 \\
\frac{f'''}{3!} & \frac{f''}{2!} & f' & f & . & . & 0 \\
. & . & . & . & . & . & . \\
. & . & . & . & . & . & . \\
\frac{f^{(n)}}{n!} & \frac{f^{(n-1)}}{(n-1)!} & . & . & . & . & f'
\end{bmatrix},
\tag{1}
$$

and let $\Delta_n = \det(M_n)$.

1°) Calculate Δ_1 and Δ_2.

2°) Prove that, for $n \geqslant 1$,

$$
\Delta_{n+1} = f' \times \Delta_n - \frac{1}{n+1} f \times \Delta_n'.
\tag{2}
$$

3°) Suppose that f never vanishes. Derive from (2) an expression for Δ_n in a closed form.

---

166. ★★ *Symmetric matrices whose k-th power is the identity*

Let $A \in \mathcal{M}_n(\mathbb{R})$ be symmetric. We suppose that there exists a positive integer k such that $A^k = I_n$.

1°) Show that, necessarily, $A^2 = I_n$.

2°) Show that if k is odd then $A = I_n$.

---

167. ★★ *Bounding the square roots of I_n*

For a positive integer n, let

$$
X_n = \left\{ A \in \mathcal{M}_n(\mathbb{R}) \mid A^2 = I_n \right\}.
$$

Is the set X_n closed? Is it bounded ?

---

168. ★★ *Commuting with all the permutation matrices*

A matrix $M_\sigma = [m_{ij}] \in \mathcal{M}_n(\mathbb{R})$ is called a permutation matrix when

$$
\begin{cases}
m_{ij} = 1 \text{ if } \sigma(j) = i, \\
m_{ij} = 0 \text{ if } \sigma(j) \neq i,
\end{cases}
$$

where $\sigma : \{1, 2, \dots, n\} \to \{1, 2, \dots, n\}$ is a permutation of $\{1, 2, \dots, n\}$.

1°) How many permutations are there?

2°) Determine the matrices $A \in \mathcal{M}_n(\mathbb{R})$ which commute with all the permutation matrices M_σ.

---

169. ★★ *Finding the determinant of a matrix decomposed by blocks (1)*

Let A, B, C, D be in $\mathcal{M}_n(\mathbb{R})$ satisfying $CD^T + DC^T = 0$. We consider the $2n \times 2n$ matrix M built up by blocks from A, B, C, D as below:

$$M = \begin{bmatrix} A & B \\ C & D \end{bmatrix}.$$

1°) (a) We firstly suppose that D is invertible; then show that

$$\det M = \det(AD^T + BC^T). \tag{1}$$

(b) With the help of a counterexample, show that (1) may fail if D is not assumed invertible.

2°) Prove that, for any D,

$$(\det M)^2 = \left[\det(AD^T + BC^T)\right]^2. \tag{2}$$

---

170. ★★ *Characteristic polynomials of permuted products of matrices*

If A and B are two (square) matrices, a standard result in matrix theory states that the characteristic polynomials of AB and BA are the same. Here we generalize this result. Consider p matrices $A_1, A_2, \dots, A_p$ in $\mathcal{M}_n(\mathbb{C})$; we set:

$$\begin{aligned} B_1 &= A_1 A_2 \dots A_p \\ B_2 &= A_2 A_3 \dots A_p A_1 \\ B_3 &= A_3 A_4 \dots A_p A_1 A_2 \\ &\dots \\ B_p &= A_p A_1 \dots A_{p-1}. \end{aligned}$$

Prove that the p matrices $B_1, B_2, \dots, B_p$ have the same characteristic polynomial.

---

171. ★★ *Expressing $AB + BA$ in terms of traces*

Let A and B be two matrices in $\mathcal{M}_2(\mathbb{C})$ whose determinants are equal to 1.

Prove the following decomposition of $AB + BA$:

$$AB + BA = \operatorname{tr}(AB)\, I_2 - \operatorname{tr}(A)\operatorname{tr}(B)\, I_2 + \operatorname{tr}(B)\, A + \operatorname{tr}(A)\, B. \tag{1}$$

---

172. ★★ *An elegant formula on traces*

Let $A \in \mathcal{M}_n(\mathbb{R})$ be such that both A and $A + I_n$ are invertible.

1°) Show that $A^{-1} + I_n$ is also invertible.

2°) Prove the following formula on traces:

$$\text{tr}\left[(A + I_n)^{-1}\right] + \text{tr}\left[(A^{-1} + I_n)^{-1}\right] = n. \tag{1}$$

---

173. ★★ *On the exponential of matrices*

1°) Let $A \in \mathcal{M}_n(\mathbb{R})$ be antisymmetric. Show that $\exp(A)$ is orthogonal with a determinant equal to 1 (it is thus a so-called rotation in $\mathbb{R}^n$).

2°) Let $A \in \mathcal{M}_n(\mathbb{R})$ be symmetric. Show that $\exp(A)$ is symmetric positive definite.

3°) Let $B \in \mathcal{M}_n(\mathbb{R})$ be symmetric positive definite. Show that there is a unique symmetric $A \in \mathcal{M}_n(\mathbb{R})$ such that $\exp(A) = B$

4°) Consider the following 2×2 matrix:

$$C = \begin{bmatrix} 0 & 2\pi \\ -2\pi & 0 \end{bmatrix}.$$

Calculate $\exp(C)$. Is the obtained result in contradiction with the previous results?

---

174. ★★ *Exponential of a 3×3 antisymmetric matrix*

Let A be the following 3×3 antisymmetric matrix:

$$A = \begin{bmatrix} 0 & -c & b \\ c & 0 & -a \\ -b & a & 0 \end{bmatrix},$$

where a, b, c are real numbers, not all zero.

Prove that:

$$\exp(A) = I_3 + \frac{\sin r}{r} A + \frac{1 - \cos r}{r^2} A^2, \tag{1}$$

where $r = \sqrt{a^2 + b^2 + c^2}$.

---

175. ★★ *Limits of quadratic forms*

Let (q_k) be a sequence of quadratic forms on $\mathbb{R}^n$:

$$\text{for all } x \text{ in } \mathbb{R}^n, \ q_k(x) = x^T A_k x,$$

where $A_k \in \mathcal{M}_n(\mathbb{R})$ is symmetric.

We suppose that the sequence of functions (q_k) converges pointwise to some limit function f. Question: is f a quadratic form on $\mathbb{R}^n$?

45

176. ★★ *Two particular quadratic forms*

The following quadratic forms on $\mathbb{R}^n$, $n \geqslant 2$, appear in some problems in Applied mathematics (Combinatorics, Optimization):

$$q_1(x_1, x_2, ..., x_n) = \sum_{1 \leqslant i < j \leqslant n} x_i x_j \quad (= x^T Q_1 x);$$

$$q_2(x_1, x_2, ..., x_n) = \sum_{i=1}^{n} x_i^2 + \sum_{1 \leqslant i < j \leqslant n} x_i x_j \quad (= x^T Q_2 x).$$

Decomposing $q_1(x_1, x_2, ..., x_n)$ as a sum of independent squares is not so easy... However, a natural question is: what is the inertia of the quadratic form q_1 (or of the associated symmetric matrix Q_1)?

Consider the same question for $q_2(x_1, x_2, ..., x_n)$ (things are easier here).

---

177. ★★ *Inequalities on sums and products of quadratic forms*

Let $A \in \mathcal{M}_n(\mathbb{R})$ be symmetric positive definite. Show the following:

$$\text{for all } x, y \text{ in } \mathbb{R}^n, \ \left(x^T A x\right) \cdot \left(y^T A^{-1} y\right) \geqslant \left(x^T y\right)^2 ; \tag{1}$$

$$\text{for all } y \neq 0 \text{ in } \mathbb{R}^n, \ \frac{1}{y^T A^{-1} y} = \min_{x^T y \neq 0} \left[\frac{x^T A x}{(x^T y)^2} \right] ; \tag{2}$$

$$\text{for all } x \text{ in } \mathbb{R}^n, \ \left(x^T A x\right) + \left(y^T A^{-1} y\right) \geqslant 2 \left(x^T y\right) . \tag{3}$$

---

178. ★★ *An improved* CAUCHY–SCHWARZ *inequality*

Let $\mathbb{R}^n$ be equipped with the usual inner-product and the associated Euclidean norm. Let x and y be two non-null vectors in $\mathbb{R}^n$.

1°) Prove the following inequality:

$$\left\| \frac{x}{\|x\|} - \frac{y}{\|y\|} \right\| \leqslant \frac{2}{\|x\| + \|y\|} \, \|x - y\| . \tag{1}$$

2°) Deduce from the above:

$$1 - 2 \left(\frac{\|x - y\|}{\|x\| + \|y\|} \right)^2 \leqslant \frac{x^T y}{\|x\| \cdot \|y\|} \leqslant -1 + 2 \left(\frac{\|x + y\|}{\|x\| + \|y\|} \right)^2 . \tag{2}$$

---

179. ★★ *Combining two partially positive semidefinite matrices*

Let A and B be two symmetric real $n \times n$ matrices, and let S and T be two subsets of $\mathbb{R}^n$. We assume the following properties:

$$
\begin{aligned}
x^T A x &\geqslant 0 \text{ for all } x \in S, \\
x^T B x &\geqslant 0 \text{ for all } x \in T, \\
S \cup T &= \mathbb{R}^n.
\end{aligned}
$$

Prove that there exists a $\lambda \in [0,1]$ such that the matrix $\lambda A + (1 - \lambda)B$ is positive semidefinite.

---

180. ★★ *Positivity of a quadratic form on a subspace (1)*

Let $Q \in \mathcal{S}_p(\mathbb{R})$ and $A \in \mathcal{M}_{m,p}(\mathbb{R})$.

Prove the equivalence of the following two statements:

(a) $x^T Q x > 0$ for all $x \neq 0$ in the kernel of A;

(b) there exists an $\alpha \geqslant 0$ such that $Q + \alpha A^T A$ is positive semidefinite.

---

181. ★★ *Positivity of a quadratic form on a subspace (2)*

Let A and B be two matrices in $\mathcal{S}_n(\mathbb{R})$. We suppose that B is not negative definite.

Prove the equivalence of the following two statements:

(a) $\left(x^T B x \geqslant 0 \text{ and } x \neq 0\right) \Longrightarrow \left(x^T A x > 0\right)$;

(b) there exists a $\mu \geqslant 0$ such that $A - \mu B$ is positive definite.

---

182. ★★ *Characterization of the everywhere positivity of a quadratic function*

Let $A \in \mathcal{S}_n(\mathbb{R})$, $b \in \mathbb{R}^n$, $c \in \mathbb{R}$, and $q : x \in \mathbb{R}^n \mapsto q(x) = x^T A x + 2b^T x + c$ be the quadratic function on $\mathbb{R}^n$ associated with these data.

One then defines the quadratic form $\widehat{q}$ on $\mathbb{R}^{n+1}$, associated with q, in the following way:

$$
(x, t) \in \mathbb{R}^n \times \mathbb{R} \mapsto \widehat{q}(x, t) = x^T A x + \left(2b^T x\right) t + ct^2. \tag{1}
$$

1°) Show that $\widehat{q}$ is a positive semidefinite quadratic form on $\mathbb{R}^{n+1}$ if and only if $q \geqslant 0$ on $\mathbb{R}^n$.

2°) Deduce the equivalence below:

$$
\left(\begin{array}{c} x^T A x + 2b^T x + c \geqslant 0 \\ \text{for all } x \in \mathbb{R}^n \end{array} \right) \Leftrightarrow \left(\widehat{A} = \begin{bmatrix} c & b^T \\ b & A \end{bmatrix} \succcurlyeq 0 \right). \tag{2}
$$

47

183. ★★ *A quadratic form on $\mathcal{M}_n(\mathbb{R})$ in terms of traces*

Let $q : \mathcal{M}_n(\mathbb{R}) \to \mathbb{R}$ be the function defined by:

$$A \in \mathcal{M}_n(\mathbb{R}) \mapsto q(A) = \mathrm{tr}(A^2).$$

Show that q is a quadratic form on $\mathcal{M}_n(\mathbb{R})$ and determine its signature.

---

184. ★★ *A sufficient condition for positive definiteness of A using an auxiliary positive definite B*

Let A and B be two symmetric real (n, n) matrices.

Prove that

$$\left(\begin{array}{c} B \text{ positive definite} \\ \text{and} \\ AB + BA \text{ positive definite} \end{array} \right) \Rightarrow (A \text{ positive definite}). \tag{1}$$

---

185. ★★ *An original characterization of positive definiteness*

Let $A \in \mathcal{M}_n(\mathbb{R})$ be symmetric and invertible. Let H denote a vector space of $\mathbb{R}^n$, not reduced to the extreme cases $\{0\}$ or $\mathbb{R}^n$ (to avoid trivialities). The orthogonal vector space of H is denoted by $H^\perp$.

1°) Show that A is positive definite if and only if:

$$\left\{ \begin{array}{c} x^T A x > 0 \text{ for all } x \neq 0 \text{ in } H \\ \text{and} \\ x^T A^{-1} x > 0 \text{ for all } x \neq 0 \text{ in } H^\perp. \end{array} \right. \tag{1}$$

2°) With the help of a counterexample, show that the result does not hold true when (1) is replaced by:

$$\left\{ \begin{array}{c} x^T A x > 0 \text{ for all } x \neq 0 \text{ in } H \\ \text{and} \\ x^T A x > 0 \text{ for all } x \neq 0 \text{ in } H^\perp. \end{array} \right. \tag{2}$$

---

186. ★★ *A "Product " of positive semidefinite matrices*

Let A and $B \in \mathcal{M}_n(\mathbb{R})$ be symmetric positive semidefinite. We define a new symmetric matrix $C = [c_{i,j}] \in \mathcal{M}_n(\mathbb{R})$ as follows:

$$c_{i,j} = a_{i,j} \times b_{i,j} \text{ for all } i, j.$$

1°) Is C positive semidefinite?

2°) What if both A and B are positive definite?

48

187. ★★ *Parallel addition of symmetric positive definite matrices*

Let A and B be two $n \times n$ symmetric positive definite matrices. We define a "mixture" of the two quadratic forms associated with A and B, in the following manner:

$$x \in \mathbb{R}^n \mapsto q_{A,B}(x) = \inf_{u+v=x} \left[u^T A u + v^T A v \right]. \tag{1}$$

1°) Show that, for every $x \in \mathbb{R}^n$, the infimum in the right-hand side of the definition (1) is finite, and that there exists a unique pair $(\overline{u}, \overline{v}) \in \mathbb{R}^n \times \mathbb{R}^n$ for which:

$$\overline{u} + \overline{v} = x, \tag{2-a}$$
$$q_{A,B}(x) = \overline{u}^T A \overline{u} + \overline{v}^T A \overline{v}. \tag{2-b}$$

2°) Show that there is an $n \times n$ symmetric positive definite matrix $\mathcal{A}$, to be expressed in terms of A and B, such that

$$q_{A,B}(x) = x^T \mathcal{A} x \text{ for all } x \in \mathbb{R}^n. \tag{3}$$

$\mathcal{A}$ is called the *parallel sum* of A and B; hereafter it will be denoted by $A//B$.

3°) We set $n = 2$. Prove that

$$A//B = \frac{\det B}{\det(A+B)} A + \frac{\det A}{\det(A+B)} B. \tag{4}$$

4°) For $x \in \mathbb{R}^n$, consider the unique pair $(\overline{u}, \overline{v})$ for which (2-a) and (2-b) hold. Prove the following relationship:

$$(A//B)\, x = A\overline{u} = B\overline{v}. \tag{5}$$

---

188. ★★ *Random convex quadratic forms*

Let $A(\omega) \in \mathcal{M}_n(\mathbb{R})$ be a random matrix such that $A(\omega)$ is almost surely symmetric positive definite ($(\Omega, \mathcal{A}, P)$ is the underlying probability space). It is assumed that both A and A^{-1} are integrable (with respect to the probability measure P).

1°) Check that the matrix $\mathbb{E}(A) = \int_\Omega A(\omega) dP$ is symmetric positive definite.

2°) Given $x \in \mathbb{R}^n$, we define $L^1(x)$ as the set of all $\mathbb{R}^n$-valued random variables X whose expectation is x (that is to say, $\mathbb{E}(X) = x$).

Find

$$\inf \left\{ \mathbb{E}(X^T A^{-1} X) : X \in L^1(x) \right\} \tag{1}$$

and the unique $X^* \in L^1(x)$ where this infimum is achieved.

---

189. ★★ *A characterization of diagonalizable real matrices*

Let $A \in \mathcal{M}_n(\mathbb{R})$. Prove that A is diagonalizable if and only if there exists a symmetric positive definite matrix S such that

$$A^T = S^{-1}AS.$$

---

190. ★★ *Expressing $x^T A^{-1}x$ with the help of determinants (1)*

Let $A \in \mathcal{M}_n(\mathbb{R})$ be invertible and let $x \in \mathbb{R}^n$.

Prove that

$$x^T A^{-1}x = \frac{\det(A + xx^T)}{\det A} - 1. \tag{1}$$

---

191. ★★ *Expressing $x^T A^{-1}x$ with the help of determinants (2)*

Let $A = [a_{i,j}] \in \mathcal{M}_n(\mathbb{R})$ be symmetric and invertible, and let $x \in \mathbb{R}^n$. We consider the enlarged symmetric matrix A_x in $\mathcal{M}_{n+1}(\mathbb{R})$ defined as follows:

$$A_x = \begin{bmatrix} a_{11} & . & . & a_{1n} & x_1 \\ . & . & . & . & . \\ . & . & . & . & . \\ a_{n1} & . & . & a_{nn} & x_n \\ x_1 & . & . & x_n & 1 \end{bmatrix}.$$

Prove that

$$x^T A^{-1}x = 1 - \frac{\det(A_x)}{\det A}. \tag{1}$$

---

192. ★★ *A "determinantal" characterization of equality of two matrices*

1°) Let $A \in \mathcal{M}_n(\mathbb{R})$ satisfy:

$$\det(A + X) = \det(X) \text{ for all } X \text{ in } \mathcal{M}_n(\mathbb{R}). \tag{1}$$

Prove that A is necessarily the null matrix.

2°) Let $A, B \in \mathcal{M}_n(\mathbb{R})$. Show that

$$(A = B) \Leftrightarrow (\det(A + X) = \det(B + X) \text{ for all } X \text{ in } \mathcal{M}_n(\mathbb{R})). \tag{2}$$

---

193. ★★ *A "trace" characterization of $AB = 0$ for two symmetric matrices A and B*

Let A and B be two real symmetric $n \times n$ matrices.

Prove the equivalence of the following two statements:

$$AB = 0; \tag{1}$$

$$\mathrm{tr}(A + B)^k = \mathrm{tr}(A^k) + \mathrm{tr}(B^k) \text{ for all positive integer } k. \tag{2}$$

---

194. ★★ *A characterization of $AB = 0$ when A is positive semidefinite*

Let $A \in \mathcal{M}_n(\mathbb{R})$ be symmetric positive semidefinite and let $B \in \mathcal{M}_n(\mathbb{R})$. We suppose that $AB + BA = 0$.

Prove that

$$AB = BA = 0.$$

---

195. ★★ *A characterization of commutativity by diagonalization*

1°) Let A and B be two matrices in $\mathcal{M}_n(\mathbb{C})$.

(a) Suppose that both A and B are diagonalizable, and that $AB = BA$. Show that A and B are simultaneously diagonalizable, that is to say: there exists an invertible matrix P such that both $P^{-1}AP$ and $P^{-1}BP$ are diagonal.

(b) Generalization. Let $(A_i)_{i \in I}$ be a collection of diagonalizable matrices such that:

$$A_i A_j = A_j A_i \text{ for all } i, j \text{ in } I. \tag{1}$$

Show that all the A_i' s are simultaneously diagonalizable.

2°) (a) Let A and B be two symmetric matrices in $\mathcal{M}_n(\mathbb{R})$ satisfying $AB = BA$. Show that there exists an orthogonal matrix U which simultaneously diagonalizes A and B.

(b) Generalization. Let $(A_i)_{i \in I}$ be a collection of real symmetric matrices such that: $A_i A_j = A_j A_i$ for all i, j in I (same condition as in (1)). Show that all the A_i' s are diagonalizable with the help of the same orthogonal matrix U.

---

196. ★★ *The compact convex set of 3×3 correlation matrices*

A symmetric positive semidefinite matrix $A = [a_{i,j}] \in \mathcal{M}_n(\mathbb{R})$ is called a *correlation matrix* when $a_{i,i} = 1$ for all $i = 1, 2, \ldots, n$.

1°) Show that the set $\mathcal{C}_n$ of all correlation matrices is a compact convex set of $\mathcal{S}_n(\mathbb{R})$.

2°) Characterization of matrices in $\mathcal{C}_3$. Prove that

$$A = \begin{bmatrix} 1 & a & c \\ a & 1 & b \\ c & b & 1 \end{bmatrix}$$

is a correlation matrix if and only if:

$$a, b, c \in [-1, 1] \text{ and } a^2 + b^2 + c^2 \leqslant 1 + 2abc. \tag{1}$$

---

197. ★★ *A maximization problem involving positive definite quadratic forms (1)*

Let $A \in \mathcal{M}_n(\mathbb{R})$ be symmetric positive definite. The following optimization problem arises in Statistics: we want to maximize

$$f(x) = x^T A^2 x - \left(x^T A x\right)^2$$

over the unit sphere of $\mathbb{R}^n$ (for the usual Euclidean norm).

What is the maximal value in this problem?

---

198. ★★ *A maximization problem involving positive definite quadratic forms (2)*

Let $A \in \mathcal{M}_n(\mathbb{R})$ be symmetric positive definite. The following optimization problem arises in Statistics: we want to maximize

$$f(x) = x^T A x - \frac{1}{x^T A^{-1} x}$$

over the unit sphere of $\mathbb{R}^n$ (for the usual Euclidean norm).

1°) What is the maximal value in this problem?

2°) At what points of the unit sphere is this maximal value attained?

---

199. ★★ *Inequalities from eigenvalue analysis of real symmetric matrices*

Let $x_1, x_2, \ldots, x_n$ be real numbers. We want to prove the following inequality:

$$\sum_{i=0}^{n} (x_i - x_{i+1})^2 \geqslant \left[4 \sin^2\left(\frac{\pi}{2(n+1)}\right)\right] \sum_{i=0}^{n} x_i^2, \tag{1}$$

where $x_0 = 0$ and $x_{n+1} = 0$ by convention.

For that purpose, we suggest the approach below.

1°) Consider the $n \times n$ symmetric matrix

$$M = \begin{bmatrix} 2 & -1 & 0 & & & & & \\ -1 & 2 & -1 & 0 & & & & \\ & -1 & 2 & -1 & 0 & & & \\ & & & \cdot & & & & \\ & & & & \cdot & & & \\ & & & & & -1 & 2 & -1 & 0 \\ & & & & & & -1 & 2 & -1 \\ & & & & & & & -1 & 2 \end{bmatrix} \cdot$$

Check that the eigenvalues of M are

$$\lambda_k = 2\left(1 - \cos\frac{k\pi}{n+1}\right) = 4\sin^2\left(\frac{k\pi}{2(n+1)}\right), \text{ for } k = 1, 2, \dots, n. \qquad (2)$$

$2°$) Prove (again) that

$$x^T M x \geqslant \lambda_{\min} \|x\|^2 \text{ for all } x \in \mathbb{R}^n, \qquad (3)$$

where $\lambda_{\min}$ refers to the smallest eigenvalue of M.

$3°$) Spot the smallest eigenvalue in (2) and conclude with the inequality (3).

---

200. ★★ *The* KANTOROVITCH *inequality in brief*

Let $A \in \mathcal{M}_n(\mathbb{R})$ be symmetric and positive definite. We intend to give a short proof of the so-called KANTOROVITCH inequality: for any unit vector $x \in \mathbb{R}^n$,

$$\left(x^T A x\right) \times \left(x^T A^{-1} x\right) \leqslant \frac{1}{4}\left(\sqrt{\frac{\lambda_1}{\lambda_n}} + \sqrt{\frac{\lambda_n}{\lambda_1}}\right)^2, \qquad (1)$$

where λ_1 (resp. λ_n) denotes the smallest (resp. the largest) eigenvalue of A.

For that purpose, we advocate following the method below:

- Firstly, diagonalize A (and A^{-1}) with an orthogonal matrix;

- Next, determine the eigenvalues of the matrix $\frac{A}{t} + tA^{-1}$, where t is specifically chosen equal to $\sqrt{\lambda_1 \lambda_n}$;

- Study the function $\lambda > 0 \mapsto \frac{\lambda}{t} + \frac{t}{\lambda}$, and conclude with the inequality $\sqrt{uv} \leqslant \frac{u+v}{2}$.

---

201. ★★ *Reducing a collection of symmetric matrices with* HELMERT *orthogonal matrices*

$1°$) Consider the following matrix in $\mathcal{M}_n(\mathbb{R})$, $n \geqslant 2$:

$$H = \begin{bmatrix} \frac{1}{\sqrt{n}} & \frac{1}{\sqrt{1\cdot2}} & \frac{1}{\sqrt{2\cdot3}} & \cdots & \frac{1}{\sqrt{(n-1)\cdot n}} \\ \frac{1}{\sqrt{n}} & -\frac{1}{\sqrt{1\cdot2}} & \frac{1}{\sqrt{2\cdot3}} & \cdots & \frac{1}{\sqrt{(n-1)\cdot n}} \\ \frac{1}{\sqrt{n}} & 0 & -\frac{2}{\sqrt{2\cdot3}} & \cdots & \frac{1}{\sqrt{(n-1)\cdot n}} \\ \cdots & \cdots & \cdots & \cdots & \cdots \\ \frac{1}{\sqrt{n}} & 0 & 0 & \cdots & -\frac{n-1}{\sqrt{(n-1)\cdot n}} \end{bmatrix}.$$

Check that H is an orthogonal matrix. It is called a HELMERT matrix.

$2°$) Consider the following collection of symmetric matrices $A_{a,b} = [a_{i,j}]$ in $\mathcal{M}_n(\mathbb{R})$, $n \geqslant 2$, with

$$a_{i,j} = a \text{ if } i = j, \ a_{i,j} = b \text{ if } i \neq j,$$

53

where a and b are real numbers.

What are the eigenvalues of $A_{a,b}$? Use a HELMERT matrix to diagonalize $A_{a,b}$.

---

202. ★★ *Expansion of* $\det(I_p + zA)$ *as the sum of a power series*

Let $A \in \mathcal{M}_p(\mathbb{C})$. We consider the power series whose general term is $\frac{(-1)^n}{n}\mathrm{tr}(A^n)\,z^n$, $z \in \mathbb{C}$.

1°) Show that the power series in question is absolutely convergent

- for $|z|$ small enough when A is not nilpotent;

- for all $z \in \mathbb{C}$ when A is nilpotent.

2°) Check that $\exp\left[\ln(1+z)\right] = 1 + z$ for all complex numbers z satisfying $|z| < 1$.

3°) Show that, for $z \in \mathbb{C}$ in a neighborhood of 0, we have the following:

$$\det(I_p + zA) = \exp\left[\sum_{n=1}^{+\infty} \frac{(-1)^{n-1}}{n}\mathrm{tr}(A^n)\,z^n\right]. \tag{1}$$

---

203. ★★ *Making equal the diagonal entries of a Hermitian matrix*

Let $A \in \mathcal{M}_n(\mathbb{C})$ be a Hermitian matrix. Prove that one can find a unitary matrix U such that $U^{-1}AU \;(= U^*AU)$ has main diagonal entries all equal to $\frac{1}{n}\mathrm{tr}(A)$.

---

204. ★★ *More on* 2×2 *orthogonal matrices*

We equip $\mathcal{M}_2(\mathbb{R})$ with the inner product $\langle U, V \rangle = \mathrm{tr}(U^T V)$ and the resulting norm $\|M\| = \sqrt{\langle M, M \rangle}$. Let us denote by $\mathcal{O}_2^+(\mathbb{R})$ (resp. $\mathcal{O}_2^-(\mathbb{R})$) the set of 2×2 orthogonal matrices whose determinant is equal to 1 (resp. equal to -1).

1°) Consider the two-dimensional vector space (or plane) P_+ of real matrices of the form $\begin{bmatrix} a & -b \\ b & a \end{bmatrix}$ and the two-dimensional vector space (or plane) P_- of real matrices of the form $\begin{bmatrix} a & b \\ b & -a \end{bmatrix}$.

- Show that P_+ and P_- are orthogonal.

- Check that $\mathcal{O}_2^+(\mathbb{R})$ (resp. $\mathcal{O}_2^-(\mathbb{R})$) is the intersection of P_+ (resp. of P_-) with the sphere of $\mathcal{M}_2(\mathbb{R})$ centered at 0 and of radius $\sqrt{2}$.

2°) Consider arbitrary $A \in \mathcal{O}_2^+(\mathbb{R})$ and $B \in \mathcal{O}_2^-(\mathbb{R})$. Show that their mutual distance $\|A - B\|$ is constantly equal to 2.

205. ★★ *Parametrization of* 3×3 *rotation matrices*

We denote by $\mathcal{SO}_3(\mathbb{R})$ (or $\mathcal{O}_3^+(\mathbb{R})$) the set of $(3,3)$ orthogonal matrices whose determinant is equal to 1. Such matrices are called special orthogonal or rotations.

1°) Let $M \in \mathcal{SO}_3(\mathbb{R})$ admit -1 as an eigenvalue.

Show that M is similar to the diagonal matrix $diag(-1, -1, -1)$.

2°) Let $M \in \mathcal{SO}_3(\mathbb{R})$ not admit -1 as an eigenvalue.

- Prove that there exists a unique antisymmetric matrix A satisfying

$$(M + I_3)(I_3 + A) = 2I_n,$$

thus allowing the following factorization of M: $M = (I_3 - A)(I_3 + A)^{-1}$.

- Noting that any antisymmetric 3×3 matrix can be written as $\begin{bmatrix} 0 & -c & b \\ c & 0 & -a \\ -b & a & 0 \end{bmatrix}$,

with real numbers a, b, c, show that M can be parameterized as

$$M = \frac{1}{a^2 + b^2 + c^2 + 1} \begin{bmatrix} a^2 - b^2 - c^2 + 1 & 2(ab + c) & 2(ac - b) \\ 2(ab - c) & -a^2 + b^2 - c^2 + 1 & 2(bc + a) \\ 2(ac + b) & 2(bc - a) & -a^2 - b^2 + c^2 + 1 \end{bmatrix}. \tag{1}$$

---

206. ★★ *Instantaneous rotations imply rotation everywhere*

Let $F : (x, y) \in \mathbb{R}^2 \mapsto F(x, y) \in \mathbb{R}^2$ be a twice continuously differentiable mapping. We denote by $JF(x, y) \in \mathcal{M}_2(\mathbb{R})$ the Jacobian matrix of F at the point (x, y). We suppose that $F(0, 0) = (0, 0)$ and

$$JF(x, y) \text{ is a rotation matrix for any } (x, y) \in \mathbb{R}^2. \tag{1}$$

Recall that a rotation matrix is an orthogonal matrix whose determinant equals 1.

Prove that F itself is a rotation (*i.e.*, there is a rotation matrix U such that $F(x, y) = U \begin{pmatrix} x \\ y \end{pmatrix}$ for all $(x, y) \in \mathbb{R}^2$).

---

207. ★★ *Impossible convex combinations of orthogonal matrices*

Let A and B be two orthogonal matrices of $\mathcal{M}_n(\mathbb{R})$ and let p and q be two positive real numbers satisfying $p + q = 1$.

We suppose that $pA + qB$ is orthogonal. What can be said about A and B?

---

208. ★★ *An "intermediate" norm on $\mathbb{R}^n$*

For $x = (x_1, x_2, \ldots, x_n) \in \mathbb{R}^n$ and an integer k between 1 and n, one defines

$$\|x\|_{(k)} = \text{sum of the } k \text{ largest elements among } |x_1|, |x_2|, \ldots, |x_n|.$$

For example, in $\mathbb{R}^3$, for $k = 2$,

$$\|(3, -2, 5)\|_{(2)} = 8; \quad \|(2, 2, 1)\|_{(2)} = 4.$$

1°) Show that $\|\cdot\|_{(k)}$ defines a norm on $\mathbb{R}^n$.

2°) Compare $\|\cdot\|_{(k)}$ with the usual norms $\|\cdot\|_\infty$ and $\|\cdot\|_1$ in $\mathbb{R}^n$. What can be deduced concerning the corresponding unit balls?

3°) One considers here $\mathbb{R}^3$ and one chooses $k = 2$. Draw, as neatly as possible,

$$S = \left\{ (x, y, z) \in \mathbb{R}^3; \; x \geqslant 0, y \geqslant 0, z \geqslant 0, \|(x, y, z)\|_{(2)} = 1 \right\}.$$

---

209. ★★ MIZUNO's *inequalities*

Let u and v in $\mathbb{R}^n$ such that $u^T v \geqslant 0$. Let $u \cdot v$ denote the vector in $\mathbb{R}^n$ whose components are $u_1 v_1, u_2 v_2, \ldots, u_n v_n$.

1°) Prove the following inequality:

$$\|u \cdot v\| \leqslant \frac{\sqrt{2}}{4} \|u + v\|^2, \tag{1}$$

where $\|\cdot\|$ stands for the usual Euclidean norm in $\mathbb{R}^n$.

2°) We suppose here that u and v are orthogonal. Prove that

$$\|u \cdot v\| \leqslant \frac{\sqrt{2}}{2} \|u\| \times \|v\|. \tag{2}$$

---

210. ★★ *A formula for decomposing sums of squares of norms of vectors*

Let E be a prehilbertian space ($\mathbb{R}^n$ equipped with the usual inner product, for example), let $\|\cdot\|$ denote the norm associated with the inner product in E and let $x_1, x_2, \ldots, x_n$ be n points in E.

Prove the following decomposition formula:

$$\sum_{\substack{i,j=1 \\ i<j}}^{n} \|x_i - x_j\|^2 = n \sum_{i=1}^{n} \|x_i\|^2 - \left\| \sum_{i=1}^{n} x_i \right\|^2. \tag{1}$$

---

211. ★★ *An "optimal" decomposition of a vector in $\mathbb{R}^n$*

Let $\mathbb{R}^n$ be equipped with the usual inner product and the associated Euclidean norm. Let $x = (x_1, x_2, \dots, x_n) \in \mathbb{R}^n$ be decomposed as a sum of the following type:

$$x = u + v, \text{ with } u \in (\mathbb{R}^+)^n \text{ and } v \in (\mathbb{R}^-)^n, \tag{1}$$

where $(\mathbb{R}^+)^n$ (respectively $(\mathbb{R}^-)^n$) denotes the set of vectors in $\mathbb{R}^n$ with nonnegative components (respectively nonpositive components).

1°) Prove the following equivalence

$$(u \text{ and } v, \text{ appearing in (1), are orthogonal}) \Leftrightarrow (u = x^+ \text{ and } v = x^-), \tag{2}$$

where $x^+ = (x_1^+, x_2^+, \dots, x_n^+)$ and $x^- = (x_1^-, x_2^-, \dots, x_n^-)$ are the vectors obtained by retaining, respectively, the nonnegative parts and the nonpositive parts of the components x_i of the vector x.

2°) Show the "optimality" of the decomposition (1), in the following sense:

$$(x \text{ decomposed as in (1)}) \Longrightarrow (\|u\| \leqslant \|x^+\| \text{ and } \|v\| \leqslant \|x^-\|). \tag{3}$$

---

212. ★★ *Ratio of volume to surface area for a solid*

Let S be a solid in $\mathbb{R}^3$ contained in a Euclidean ball of radius r. We assume that the boundary $bd(S)$ of S is a smooth surface.

Prove that

$$\text{vol}(S) \leqslant \frac{r}{3} \times \text{area}\,[bd(S)]. \tag{1}$$

The case of the closed ball of radius r shows that the inequality (1) is sharp.

---

213. ★★ *Range of a polynomial function*

1°) The case of one variable.

Let $P(x)$ be a polynomial function of the real variable x. Verify that the range $P(\mathbb{R})$ of $\mathbb{R}$ by P is a closed interval in $\mathbb{R}$.

2°) The case of several variables.

Let $P(x, y)$ be a polynomial function of two real variables x and y. Show with an example that the range $P(\mathbb{R}^2)$ of $\mathbb{R}^2$ by P could be an open interval in $\mathbb{R}$.

---

214. ★★ *Preserving nonnegativity of polynomials by adding their successive derivatives*

Let P be a real polynomial function of degree n which is nonnegative on $\mathbb{R}$. We define a new polynomial function Q as follows:

$$Q = P + P' + \ldots + P^{(n)}.$$

Prove that Q is again nonnegative on $\mathbb{R}$.

---

215. ★★ *Roots of a polynomial vs roots of its derivative*

1°) The case of 2 roots.

Consider the polynomial function $P(x) = (x - r_1)(x - r_2)$, with two roots (possibly complex) r_1 and r_2. What is the root of the derivative polynomial P'?

2°) The case of 3 roots.

Consider the polynomial function $P(x) = (x - r_1)(x - r_2)(x - r_3)$, with three roots r_1, r_2 and r_3. We suppose that the three roots are distinct and non-aligned, so that they form a triangle $\mathcal{T}$ in the complex plane. Check on some examples that the two roots of the derivative polynomial P' lie inside the triangle $\mathcal{T}$.

3°) The general case of n roots.

Consider a non-constant polynomial function P, with complex coefficients. Prove that the roots of the derivative polynomial P' lie in the convex hull of the roots of P (*i.e.*, in the smallest convex polygon containing the roots of P).

---

216. ★★ *A continuous function takes equal values at some opposite points on a circle*

We consider $\mathbb{R}^2$ equipped with the usual Euclidean norm and we designate by S the unit sphere of $\mathbb{R}^2$.

1°) Show that S is a connected compact set.

2°) Let f be a real-valued continuous function on S. Prove that there necessarily exist two diametrically opposed points on S at which f takes equal values.

---

217. ★★ *Local minimizers vs global minimizers*

Let $f : \mathbb{R}^n \to \mathbb{R}$ be a continuous function and $\overline{x} \in \mathbb{R}^n$. Prove the following equivalence:

$$\left(\begin{array}{c} \overline{x} \text{ is a global minimizer} \\ \text{of } f \end{array} \right) \Longleftrightarrow \left(\begin{array}{c} \text{Every } x^* \text{ such that } f(x^*) = f(\overline{x}) \\ \text{is a local minimizer of } f \end{array} \right). \tag{1}$$

---

218. ★★ *A local minimizer along all the lines but not a local minimizer on $\mathbb{R}^2$*

Let $f : \mathbb{R}^2 \to \mathbb{R}$ be defined as $f(x, y) = (y - x^2)(y - 3x^2)$.

1°) Show that $(0, 0)$ is the only critical point of f.

2°) Check that $(0, 0)$ is not a local minimizer of f on $\mathbb{R}^2$.

3°) Let d be a non-null vector of $\mathbb{R}^2$ and let $f_d : \mathbb{R} \to \mathbb{R}$ be the trace of f along the line passing through the origin and directed by d, that is to say: $f_d(t) = f(td)$.

Verify that, for all d, the point $\bar{t} = 0$ is a local minimizer of f_d.

---

219. ★★ *Functions for which all points are local minimizers*

Let $f : \mathbb{R}^n \to \mathbb{R}$ be a function. We suppose that all the points $x \in \mathbb{R}^n$ are local minimizers of f.

1°) The differentiable case.

We assume that f is a differentiable function. Is f a constant function?

2°) The continuous case.

We assume that f is a continuous function. Is f a constant function?

3°) The lower-semicontinuous case.

We assume that f is a lower-semicontinuous function. Is f a constant function?

4°) We now suppose that all points $x \in \mathbb{R}^n$ are global minimizers of f. What can be said about f?

---

220. ★★ *A necessary condition for approximate optimality*

Let $f : \mathbb{R}^n \to \mathbb{R}$ be a function assumed continuous and bounded from below on $\mathbb{R}^n$. Let $\varepsilon > 0$ and consider an ε-minimizer u_ε of f, that is to say a point satisfying

$$f(u_\varepsilon) \leqslant \inf_{x \in \mathbb{R}^n} f(x) + \varepsilon.$$

Given another parameter $\lambda > 0$, one considers a perturbed version g of the function f:

$$g : x \in \mathbb{R}^n \mapsto g(x) = f(x) + \frac{\varepsilon}{\lambda} \|x - u_\varepsilon\|.$$

1°) Show that there exists a $v \in \mathbb{R}^n$ minimizing g on $\mathbb{R}^n$:

$$g(v) = \min_{x \in \mathbb{R}^n} g(x).$$

2°) Show that v satisfies the following three properties:

$$f(v) \ \leqslant \ f(u_\varepsilon); \tag{1}$$
$$\|v - u_\varepsilon\| \ \leqslant \ \lambda; \tag{2}$$
$$f(v) \ \leqslant \ f(x) + \frac{\varepsilon}{\lambda} \|x - v\| \ \text{for all } x \in \mathbb{R}^n. \tag{3}$$

3°) An application. Suppose that $f : \mathbb{R}^n \to \mathbb{R}$ is differentiable and bounded from below on $\mathbb{R}^n$. Show that, for any $\varepsilon > 0$, there exists an $x_\varepsilon \in \mathbb{R}^n$ such that

$$\|\nabla f(x_\varepsilon)\| \leqslant \varepsilon. \tag{4}$$

---

221. ★★ *An identity on the inverse of the hyperbolic tangent function*

Let x, y, z lie in the interval $(-1, 1)$.

1°) Show that $\frac{x+y+z+xyz}{1+xy+xz+yz}$ also lies in $(-1, 1)$.

2°) Establish the following identity:

$$\operatorname{arctanh}(x) + \operatorname{arctanh}(y) + \operatorname{arctanh}(z) = \operatorname{arctanh}\left(\frac{x+y+z+xyz}{1+xy+xz+yz}\right). \tag{1}$$

---

222. ★★ *On real parts and imaginary parts of eigenvalues of complex matrices*

Let A and B be two $n \times n$ real symmetric matrices, and let $\mathcal{A} = A + iB \in \mathcal{M}_n(\mathbb{C})$. We set

$$K = \left\{ (x^T A x, x^T B x) \mid x \in \mathbb{R}^n, \|x\| = 1 \right\}.$$

We admit here that K is convex (to prove this, for $n \geqslant 3$, is the aim of Tapa 320).

Let $\lambda \in \mathbb{C}$ be an eigenvalue of $\mathcal{A}$.

Prove that $(\operatorname{Re}\lambda, \operatorname{Im}\lambda) \in K$.

---

223. ★★ *A convex set whose closure is the whole space*

Let C be a convex set of $\mathbb{R}^n$ such that $\overline{C} = \mathbb{R}^n$.

Describe all such sets C.

---

224. ★★ *Determining the maximal dimension of a vector space*

Given $u \in \mathbb{R}^n$ and $v \in \mathbb{R}^m$, let $\varphi_{u,v}$ denote the linear form defined on $L(\mathbb{R}^m, \mathbb{R}^n)$ by:

$$A \in L(\mathbb{R}^m, \mathbb{R}^n) \mapsto \varphi_{u,v}(A) = u^T A v. \tag{1}$$

Here, $L(\mathbb{R}^m, \mathbb{R}^n)$ stands for the vector space of linear mappings from $\mathbb{R}^m$ into $\mathbb{R}^n$. We note

$$\Phi = \left\{ \varphi_{u,v} \mid u \in \mathbb{R}^n \text{ and } v \in \mathbb{R}^m \right\}.$$

The set Φ is not a vector space. Moreover, any linear form on $L(\mathbb{R}^m, \mathbb{R}^n)$ is not of the form $\varphi_{u,v}$ (for example, the linear form $A \mapsto \operatorname{tr}(A)$ does not belong to Φ if $m = n > 1$).

So, a natural question is: what is the maximal dimension of a vector space contained in Φ?

225. ★★ *On the graph of a continuous mapping*

1°) The one-dimensional case.

Let $f : \mathbb{R} \to \mathbb{R}$ be a function whose graph $G_f = \{(x, f(x)) \mid x \in \mathbb{R}\} \subset \mathbb{R}^2$ is closed and connected. Is f a continuous function?

2°) The higher-dimensional case.

Let K be a compact subset of $\mathbb{R}^n$ and let $f : K \to \mathbb{R}^m$ be a mapping. Prove that f is continuous if and only if its graph $G_f = \{(x, f(x)) \mid x \in K\} \subset \mathbb{R}^n \times \mathbb{R}^m$ is compact.

226. ★★ *When is a convex set transformed into a convex set?*

Let $n \geqslant 2$ and let $f : \mathbb{R}^n \to \mathbb{R}^n$ be an injective mapping.

1°) Suppose that the image by f of any convex set is a convex set. Prove that f is an affine mapping.

2°) Show, with the help of a counterexample, that the result above may fail if f is not assumed to be injective.

227. ★★ *Surjectivity of 1-coercive continuous functions*

Let $f : \mathbb{R}^n \to \mathbb{R}$ be a continuous function. We suppose that f is 1-coercive on $\mathbb{R}^n$, that is to say:
$$|f(x)| \to +\infty \text{ as } \|x\| \to +\infty.$$

Prove that f is surjective (*i.e.* $f(\mathbb{R}^n) = \mathbb{R}$) if and only if $f(\mathbb{R}^n)$ is open.

228. ★★ *A strange minimization problem (1)*

Let $f : \mathbb{R}^2 \to \mathbb{R}$ be defined as follows:
$$f(x, y) = 2x^3 + 3e^{2y} - 6xe^y.$$

1°) Show that there is only one critical point $(a, b) \in \mathbb{R}^2$ of f, and that this critical point is a strict local minimizer of f.

2°) Show that f is not bounded from below on $\mathbb{R}^2$.

229. ★★ *A strange polynomial minimization problem (2)*

Let $f : \mathbb{R}^2 \to \mathbb{R}$ be the nonnegative polynomial function defined as follows:
$$f(x, y) = (xy - 1)^2 + x^2.$$

1°) Is the infimum of f on $\mathbb{R}^2$ achieved?

2°) What are the critical points of f and their status (local minimizer, local maximizer, or saddle point)?

230. ★★ *A strange polynomial minimization problem (3)*

Let $f : \mathbb{R}^2 \to \mathbb{R}$ be the nonnegative polynomial function defined as follows:

$$f(x,y) = (x^2y - x - 1)^2 + (x^2 - 1)^2.$$

Show that there are exactly two global minimizers but no other critical point.

---

231. ★★ *Example of determination of a locus*

Consider the two following sets in the plane $\mathbb{R}^2$ (equipped with a coordinate system Oxy): the unit circle $\mathcal{C}$ (with equation $x^2 + y^2 = 1$) and the (horizontal) line Ox (with equation $y = 0$).

Determine all the points P in the plane such that the (usual Euclidean) distance from P to $\mathcal{C}$ and that from P to D are equal.

---

232. ★★ *Majorizing a double integral with the help of 4-th order partial derivatives*

Let $f \in \mathcal{C}^4([0,1] \times [0,1], \mathbb{R})$ with: $f(x,y) = 0$ for all (x,y) on the boundary of the square $[0,1] \times [0,1]$. We intend to prove the following inequality

$$\left| \int_0^1 \int_0^1 f(x,y) \, dxdy \right| \leqslant \frac{1}{144} \sup_{(x,y) \in [0,1] \times [0,1]} \left| \frac{\partial^4 f}{\partial x^2 \partial y^2}(x,y) \right|. \tag{1}$$

1°) Let $\varphi \in \mathcal{C}^2([0,1]), \mathbb{R})$. Prove that .

$$\int_0^1 \varphi(t) \, dt = \frac{1}{2}[\varphi(0) + \varphi(1)] - \frac{1}{2}\int_0^1 t(1-t)\varphi''(t) \, dt. \tag{2}$$

2°) By applying the above property (2) successively to the functions

$$\varphi : y \mapsto \varphi(y) = f(x,y) \text{ (for fixed } x \in [0,1]),$$

and

$$\varphi : x \mapsto \varphi(x) = \frac{\partial^2 f}{\partial y^2}(x,y) \text{ (for fixed } y \in [0,1]),$$

and making use of the rule allowing the calculation of a double integral by evaluating two successive simple integrals, prove that

$$\int_0^1 \int_0^1 f(x,y) \, dxdy = \frac{1}{4}\int_0^1 \int_0^1 xy(1-x)(1-y)\frac{\partial^4 f}{\partial x^2 \partial y^2}(x,y) \, dxdy. \tag{3}$$

3°) Deduce from the equality (3) the inequality (1).

4°) Show that, without further assumptions on f, the coefficient $1/144$ appearing in inequality (1) cannot be improved.

---

233. ★★ *Surface area of a parachute*

The cloth of a parachute or paraglider is described by a surface Σ in $\mathbb{R}^3$ with equation:

$$x^2 + y^2 + z^2 = a^2, \ \frac{x^2}{a^2} + \frac{y^2}{b^2} \leqslant 1, \ z \geqslant 0,$$

where $0 < b \leqslant a$ are given parameters. Thus, Σ is a piece of a half-sphere delimited by an elliptic cylinder.

1°) Sketch the surface Σ.

2°) Calculate the surface area of Σ.

---

234. ★★ *Volume of a diamond-shaped solid*

A diamond-shaped solid S (in $\mathbb{R}^3$) is delimited above by the sphere centered at the origin and of radius $r = 3$ units (say centimeters), thus of equation $x^2 + y^2 + z^2 = 9$, and laterally by the vertical cone with equation $z = \sqrt{x^2 + y^2}$.

1°) Sketch the solid S.

2°) Calculate the volume of S.

3°) What proportion of the ball of radius $r = 3$ units does the solid S occupy?

---

235. ★★ *Volume of the* VIVIANI *solid*

A solid S (in $\mathbb{R}^3$) is delimited above by the sphere of equation $x^2 + y^2 + z^2 = 4r^2$, laterally by the cylinder with equation $x^2 + (y - r)^2 = r^2$, and below by the plane with equation $z = 0$. To be mathematically more precise,

$$S = \left\{ x, y, z \ : \ x^2 + y^2 + z^2 \leqslant 4r^2, \ x^2 + (y - r)^2 \leqslant r^2 \text{ and } z \geqslant 0 \right\}.$$

1°) Sketch the solid S.

2°) Calculate the volume of S.

---

236. ★★ *Surface area and volume of the intersection of two perpendicular cylinders*

1°) Sketch the part Σ of the cylinder with equation $x^2 + z^2 = r^2$ cut out by the cylinder with equation $x^2 + y^2 = r^2$.

For clear symmetry reasons, it is enough to take into consideration only the part Σ_0 of Σ with $x \geqslant 0, y \geqslant 0$ and $z \geqslant 0$.

2°) Calculate the surface area of Σ_0.

3°) Calculate the volume of the solid S defined as

$$S = \left\{ x, y, z \ : \ x^2 + y^2 \leqslant r^2, \ x^2 + z^2 \leqslant r^2 \right\}. \tag{1}$$

The solid S is shaped like a hard candy. For symmetry reasons, considering only the part S_0 of S with $x \geqslant 0, y \geqslant 0$ and $z \geqslant 0$ is sufficient.

---

237. ★★ *An oval with a constant width*

A mechanical piece used for camshafts has the form of a convex oval, the boundary of which is described by a curve Γ whose equation is:

$$x(t) = \frac{2}{3}\cos(t) - \frac{1}{3}\cos(2t) - \frac{8}{3}\cos\left(\frac{t}{2}\right),$$

$$y(t) = \frac{2}{3}\sin(t) + \frac{1}{3}\sin(2t) + \frac{8}{3}\sin\left(\frac{t}{2}\right),$$

for $t \in [0, 4\pi]$.

1°) With the help of a graph plotter, plot Γ, and place on the drawing the points of Γ corresponding to the following values of the parameter t : $(t = 0$ and $t = 2\pi)$; $(t = \pi$ and $t = 3\pi)$.

2°) We concentrate our attention on the point P of Γ corresponding to the value $\frac{\pi}{2}$ of the parameter t, as well as on the point Q of Γ corresponding to the value $\frac{5\pi}{2}$ of the parameter t.

- Calculate the coordinates of the points P and Q, and mark them on the drawing of Γ .

- Determine the equations of the tangent lines to Γ at P and at Q ; Δ_1 denotes the tangent line at P and Δ_2 the tangent line at Q.

- Observing that the lines Δ_1 and Δ_2 are parallel, calculate the distance between Δ_1 and Δ_2.

3°) Calculate the length of the curve Γ, and compare it with the perimeter of a circle of diameter PQ.

---

238. ★★ *Volume of a lake*

A lake has, on the horizontal plane, a border described by an ellipse with equation $\frac{x^2}{a^2} + \frac{y^2}{b^2} = 1$, where $a \geqslant b > 0$. A function, called an "elliptic sinusoid", generally constitutes an adequate approximation of the bottom of the lake, that is to say: this function gives, with enough precision, the depth z of the lake at any point with coordinates (x, y) on the surface of the lake. In the present case, the elliptic sinusoid whose mathematical expression is

$$z = f(x, y) = h_M \cos\left(\frac{\pi}{2}\sqrt{\frac{x^2}{a^2} + \frac{y^2}{b^2}}\right),$$

with $h_M > 0$ given, describes this depth well. Clearly, one only considers the points (x, y) such that $\frac{x^2}{a^2} + \frac{y^2}{b^2} \leqslant 1$ (those corresponding to the surface of the lake).

1°) Preliminaries.

(a) Sketch the shape of the lake in 3D, with the orientation of the Oz axis towards the bottom.

(b) At what point of the surface of the lake is the depth maximal?

(c) What is the area of the lake (at the level of the ground)?

2°) We plan to determine the volume of the lake in terms of a, b and h_M.

(a) What integral do we have to calculate?

(b) Calculate this volume.

---

239. ★★ *Volume of a sports centre*

A sports centre has the form of a solid delimited by the following surfaces:

- the horizontal plane with equation $z = 0$;

- the vertical planes with equations $x = 1, x = -1, y = 1, y = -1$;

- the upper surface with equation $x^2 - y^2 + z - 2 = 0$.

The solid (*i.e.*, the sports centre) itself is the set

$$S := \{(x, y, z) \in \mathbb{R}^3 \mid z \geq 0, \ -1 \leq x \leq 1, \ -1 \leq y \leq 1, \ x^2 - y^2 + z - 2 \leq 0\}.$$

We designate by Σ the upper surface of S (or the roof of the sports centre, of an incurved form),

$$\Sigma = \{(x, y, z) \in \mathbb{R}^3 \mid z \geq 0, \ -1 \leq x \leq 1, \ -1 \leq y \leq 1, \ x^2 - y^2 + z - 2 = 0\}.$$

The basic unit is $u = 1$ (without further precision); the numerical data do not come from a real physical construction, scalings have been performed in order to make mathematical calculations easier.

1°) (a) Sketch the sports centre S and its roof Σ.

(b) With the help of two rectangular parallelepipeds (of heights $h = 1$ and $h = 3$ respectively) squeezing S, provide a rough estimate for the volume of S.

(c) The roof Σ has to be consolidated with rafters installed at the intermediate height (from the ground) $h = 2$. For that, it is asked to determine the intersection of Σ with the plane of equation $z = 2$.

2°) Calculation of the volume of S.

(a) Give a Cartesian equation, in an explicit form, of Σ, that is to say an equation of the form $(x, y) \in D \longmapsto z = f(x, y)$, where the domain $D \subset \mathbb{R}^2$ and the function f have to be determined.

(b) Calculate the volume V of S.

3°) Calculation of a surface area.

One modifies S by taking as a basis (on the ground) the disk D of equation $x^2+y^2 \leqslant 1$, instead of the square of equation $(-1 \leq x \leq 1$ and $-1 \leq y \leq 1)$ as before; one designates by Σ_1 the new upper surface (or roof of the sports centre).

We now want to calculate the surface area $\mathcal{A}$ of Σ_1.

(a) Set the calculation $\mathcal{A}$ of Σ_1 in the form of an integral.

(b) Calculate $\mathcal{A}$.

---

240. ★★ *Volume of a grain silo*

A grain silo has the form of a solid delimited by the following surfaces:

- the horizontal plane with equation $z = 0$;

- a cylinder with equation $x^2 + y^2 = 9$;

- a sphere with equation $x^2 + y^2 + z^2 = 25$.

The solid (*i.e.*, the grain silo) itself is the set

$$S := \{(x, y, z) \in \mathbb{R}^3 \mid z \geq 0, \; x^2 + y^2 \leq 9, \; x^2 + y^2 + z^2 \leq 25\}.$$

We designate by Σ the upper surface of S (or the roof of the grain silo, a portion of a half-sphere),

$$\Sigma = \{(x, y, z) \in \mathbb{R}^3 \mid z \geq 0, \; x^2 + y^2 \leq 9, \; x^2 + y^2 + z^2 = 25\}.$$

The basic unit is $u = 1$ (without further precision); the numerical data do not come from a real physical construction, scalings have been performed in order to make mathematical calculations easier.

1°) (a) Sketch the silo S and its roof Σ.

(b) What is the height h of the inner surface of S?

2°) Calculation of the volume of S.

(a) Give a Cartesian equation, in an explicit form, of Σ, that is to say an equation of the form $(x, y) \in D \longmapsto z = f(x, y)$, where the domain $D \subset \mathbb{R}^2$ and the function f have to be determined.

(b) Calculate the volume V of S.

3°) Calculation of the area of the roof.

A roofer has to place a fine metal protection on Σ. He therefore has to know the surface area of it in order to prepare an estimate.

(a) Set the calculation $\mathcal{A}$ of Σ in the form of an integral.

(b) Calculate $\mathcal{A}$. What proportion of the area of the half-sphere does $\mathcal{A}$ represent?

241. ★★ *Length, area, and center of mass of a cycloid*

A cycloid is the curve described in the plane by a point P fixed on a circle of radius r which rolls, without sliding, on a horizontal line. The parametric coordinates of such a cycloid are given as below:

$$t \in \mathbb{R} \mapsto \overrightarrow{\gamma}(t) = \begin{cases} x(t) = r(t - \sin(t)), \\ y(t) = r(1 - \cos(t)). \end{cases}$$

One considers only an arc of the cycloid, denoted by C, corresponding to the values of t between 0 and 2π, that is to say:

$$C = \{\overrightarrow{\gamma}(t) \ : \ t \in [0, 2\pi]\}.$$

1°) With the help of a graph plotter, sketch the arc of the cycloid C.

2°) Calculate the length l of C (the units are not made explicit here and hereafter).

3°) Calculate the area $\mathcal{A}$ of the portion of the plane delimited by C and the horizontal axis.

4°) We assume that the density of mass of the wire C is constant. The objective here is to determine the center of mass $G = (x_G, y_G)$ of C.

(a) Indicate, without any further computation, what x_G should be.

(b) Calculate y_G.

---

242. ★★ *Comparing the lengths of neighboring arcs*

For $\alpha \geqslant 1$, let us consider the arcs Γ_α in the box $[0, 1] \times [0, 1]$ of the plane whose parametric coordinates are given as follows:

$$t \in \left[0, \frac{\pi}{2}\right] \mapsto \overrightarrow{\gamma}(t) = \begin{cases} x(t) = \cos^\alpha(t), \\ y(t) = \sin^\alpha(t). \end{cases}$$

We restrict our study to the arcs $\Gamma_1, \Gamma_2, \Gamma_3, \Gamma_4$.

1°) What are Γ_1 and Γ_2?

2°) Provide the Cartesian equations of $\Gamma_1, \Gamma_2, \Gamma_3, \Gamma_4$.

3°) Comparing Γ_3 and Γ_4, which one is the most "curved inwards"?

4°) Studying Γ_4 in more detail.

- Show that Γ_4 is actually a piece of a parabola.

- Calculate its length $l(\Gamma_4)$.

5°) Studying Γ_3 in more detail.

- Calculate the length $l(\Gamma_3)$ of Γ_3.

- Calculate the area $\mathcal{A}_3$ of the domain delimited by the Ox axis, the Oy axis, and Γ_3.

---

243. ★★ *Volume and center of mass of a funnel-shaped solid*

The funnel of a liner is shaped like a solid S delimited by the following surfaces:

- two horizontal planes of equations $z = z_1$ and $z = z_2$, with $z_2 > z_1 > 0$;

- a lateral surface of equation $z(x^2 + y^2) = r^2 x$, with $r > 0$.

Indeed,

$$S = \left\{(x, y, z) \in \mathbb{R}^3 : z_1 \leqslant z \leqslant z_2 \text{ and } z(x^2 + y^2) \leqslant r^2 x \right\}.$$

1°) (a) Check the symmetries of S (with respect to the axis Oy and with respect to the plane xOz).

(b) Determine the intersections of S with the plane xOz and with the plane of equation $z = h$.

(c) Use the properties described above to sketch the solid S.

2°) Calculate the volume V of S.

3°) Assuming that the solid S is homogeneous (*i.e.*, the density of mass is constant), calculate the coordinates $(x_G, 0, z_G)$ of the center of mass G of S.

---

244. ★★ *Area and center of mass of a blade*

An aeronautical piece S, a blade in fact, is described in the plane xOy as follows:

$$S = \left\{(x, y) \in \mathbb{R}^2 : x^2 \leqslant 8y \leqslant 8x^2 \text{ and } y^2 \leqslant 8x \leqslant 8y^2 \right\}.$$

1°) Sketch S.

2°) Calculate the area of S.

3°) Assuming that S is homogeneous, determine the center of mass $G = (x_G, y_G)$ of S.

---

245. ★★ *A variant of the change of variables formula*

The (chain) rule for differentiating an integral,

$$\frac{d}{dt}\left(\int_{x(t)}^{y(t)} f(t, x)\, dx\right) = \int_{x(t)}^{y(t)} \frac{\partial f}{\partial t}(t, x)\, dx + f\left[t, y(t)\right] y'(t) - f\left[t, x(t)\right] x'(t) \quad (1)$$

contains, as particular cases, the so-called fundamental theorem of Calculus

$$\frac{d}{dt}\left(\int_a^t f(x)\, dx\right) = f(t), \quad (2)$$

as well as the usual differential formula under the integral sign

$$\frac{d}{dt}\left(\int_a^b f(t, x)\, dx\right) = \int_a^b \frac{\partial f}{\partial t}(t, x)\, dx. \quad (3)$$

But it also hides another version of the change of variables formula, that we are going to present now.

Let $f : (t, x) \in [a, b] \times [c, d] \subset \mathbb{R}^2 \to f(t, x) \in \mathbb{R}$ be a continuous function with continuous partial derivative with respect to the first variable $((t, x) \mapsto \frac{\partial f}{\partial t}(t, x)$ is continuous on $[a, b] \times [c, d])$, and let $x : t \in [a, b] \mapsto x(t) \in [c, d]$ be continuously differentiable.

1°) Prove that

$$\int_a^b f\left[t, x(t)\right] x'(t) dt = \int_{x(a)}^{x(b)} f(b, x) \, dx - \int_a^b \left(\int_{x(a)}^{x(t)} \frac{\partial f}{\partial t}(t, x) \, dx \right) dt. \qquad (4)$$

2°) Check that formula (4) contains the usual change of variables, namely

$$\int_a^b f\left[x(t)\right] x'(t) dt = \int_{x(a)}^{x(b)} f(x) \, dx, \qquad (5)$$

and the integration by parts formula, that is

$$\int_a^b f(t) x'(t) dt = f(b) x(b) - f(a) x(a) - \int_a^b f'(t) x(t) dt. \qquad (6)$$

---

246. ★★ *Normal vectors to sides of a triangle add up to zero*

Consider a non-degenerate triangle ABC with $BC = a, AB = c, AC = b$. Orthogonal to each of its three sides, construct an outward normal vector whose length equals that of the corresponding side: $\overrightarrow{u}$, of length a, is pointing outside of the side BC; $\overrightarrow{v}$, of length b, is pointing outside of the side AC; $\overrightarrow{w}$, of length c, is pointing outside of the side AB.

Prove that $\overrightarrow{s} = \overrightarrow{u} + \overrightarrow{v} + \overrightarrow{w}$ is the zero vector.

---

247. ★★ *A minimization property of the area of an inscribed triangle*

Let ABC be a non-degenerate triangle. Let P, Q, R be three points lying on the sides BC, AC, AB, respectively, of ABC. Endpoints of the sides are excluded, however. Let x, y, z, w denote the areas of the triangles ARQ, BRP, CPQ, PQR, respectively.

Prove that

$$w > \min(x, y, z) \qquad (1)$$

unless P, Q, R are the mid-points of the sides of the triangle ABC, in which case:

$$w = x = y = z.$$

In such a case, PQR is called the *medial triangle* of the triangle ABC.

---

248. ★★ *A characterization by optimization of the orthocenter of a triangle*

Let $\mathcal{T} = ABC$ be an acute-angled triangle; hence its orthocenter (= meeting point of the three altitudes) lies in the interior of $\mathcal{T}$.

1°) Let $f : P \in \mathcal{T} \mapsto f(P) = \|PA\|$ (the usual Euclidean distance from P to the vertex A of $\mathcal{T}$).

What is the gradient vector of the function f at P? There are two further functions of this type, associated with the two other vertices.

2°) Let $g : P \in \mathcal{T} \mapsto g(P) = \|PA'\|$, where A' is the orthogonal projection of P onto the side BC of $\mathcal{T}$.

What is the gradient vector of the function g at P? There are two further functions of this type, associated with the two other sides.

3°) Let $h : \mathcal{T} \mapsto \mathbb{R}$ denote the following function

$$h(P) \;=\; \|PA\| + \|PB\| + \|PC\| + \\ \|PA'\| + \|PB'\| + \|PC'\|,$$

where A', B', C' denote the projections of P onto the sides of $\mathcal{T}$.

(a) Check that h is strictly convex and differentiable on $\mathcal{T}$.

(b) Show that the orthocenter of $\mathcal{T}$ minimizes h on $\mathcal{T}$, and that it is the only point inside $\mathcal{T}$ enjoying this "variational" property.

---

249. ★★ *A property of three "orthogonally moving points" on an ellipsoid*

Let $\mathcal{E}$ be an ellipsoid in $\mathbb{R}^3$ whose Cartesian equation in the orthonormal basis $(O; \overrightarrow{i}, \overrightarrow{j}, \overrightarrow{k})$ is:

$$\frac{x^2}{a^2} + \frac{y^2}{b^2} + \frac{z^2}{c^2} = 1.$$

1°) Consider the three "vertices" $A = (a, 0, 0), B = (0, b, 0), C = (0, 0, c)$ in $\mathcal{E}$. Calculate

$$\frac{1}{\left\|\overrightarrow{OA}\right\|^2} + \frac{1}{\left\|\overrightarrow{OB}\right\|^2} + \frac{1}{\left\|\overrightarrow{OC}\right\|^2}.$$

2°) Consider now three points A_1, A_2, A_3 in $\mathcal{E}$ such that the vectors $\overrightarrow{OA_1}, \overrightarrow{OA_2}, \overrightarrow{OA_3}$ are pairwise orthogonal.

(a) Show that $\frac{1}{\left\|\overrightarrow{OA_1}\right\|^2} + \frac{1}{\left\|\overrightarrow{OA_2}\right\|^2} + \frac{1}{\left\|\overrightarrow{OA_3}\right\|^2}$ is an invariant which does not depend on the specific points A_1, A_2, A_3 and which can be expressed in terms of a, b, c.

(b) Let P be the plane passing through the three points A_1, A_2, A_3. Show that P remains tangent to a sphere centered at 0 and whose radius can be expressed in terms of a, b, c.

250. ★★ PYTHAGORAS' *theorem for areas in a trirectangular tetrahedron*

Let $OABC$ be a trirectangular tetrahedron, that is to say: with three perpendicular triangular faces OAB, OAC, OBC, and one "hypotenuse-face" ABC.

Let S_1, S_2, S_3 denote the areas of the perpendicular faces and let S denote the area of the hypotenuse-face. Prove that:

$$S^2 = S_1^2 + S_2^2 + S_3^2. \tag{1}$$

---

251. ★★ *The magic of* FOURIER *series (1)* (Easy)

Let $f : \mathbb{R} \to \mathbb{R}$ be the function defined as follows:

$$x \in \mathbb{R} \mapsto f(x) = \begin{cases} \sin(x) & \text{if } \sin(x) \geqslant 0, \\ 0 & \text{if } \sin(x) < 0. \end{cases}$$

(In short, $f(x) = \max[\sin(x), 0]$.)

1°) Sketch the graph of f for x between -2π and 2π.

2°) Calculate the (real) FOURIER coefficients associated with f (denoted a_n and b_n).

3°) Give a clearly argued reason why f can be written as

$$f(x) = a_0 + \sum_{n=1}^{+\infty} [a_n \cos(nx) + b_n \sin(nx)] \text{ for all } x \in \mathbb{R}.$$

4°) By choosing a particular x in the above expansion, determine the value of $\sum_{p=1}^{+\infty} \frac{1}{4p^2-1}$.

5°) With the help of PARSEVAL's theorem, to be recalled, determine $\sum_{p=1}^{+\infty} \frac{1}{(4p^2-1)^2}$.

---

252. ★★ *The magic of* FOURIER *series (2)* (Easy)

Let $f : \mathbb{R} \to \mathbb{R}$ be a π-periodic function with

$$f(x) = x(\pi - x) \text{ for } 0 \leqslant x \leqslant \pi.$$

1°) Sketch the graph of f for x between -2π and 2π.

2°) Calculate the (real) FOURIER coefficients associated with f (denoted a_n and b_n).

3°) Give a clearly argued reason why f can be written as

$$f(x) = a_0 + \sum_{n=1}^{+\infty} [a_n \cos(nx) + b_n \sin(nx)] \text{ for all } x \in \mathbb{R}.$$

4°) By choosing a particular x in the above expansion, determine the value of $\sum_{p=1}^{+\infty} \frac{(-1)^{p+1}}{p^2}$.

5°) With the help of PARSEVAL's theorem, to be recalled, determine $\sum_{p=1}^{+\infty} \frac{1}{p^4}$.

---

253. ★★ *The magic of* FOURIER *series (3)* (Easy)

Let $f : \mathbb{R} \to \mathbb{R}$ be a 2π-periodic even function with

$$f(x) = \left\{ \begin{array}{l} 1 \text{ if } 0 \leqslant x \leqslant \frac{\pi}{2}, \\ -1 \text{ if } \frac{\pi}{2} < x \leqslant \pi. \end{array} \right.$$

1°) Sketch the graph of f for x between $-\pi$ and π.

2°) Calculate the (real) FOURIER coefficients associated with f (denoted a_n and b_n).

3°) Give a clearly argued reason why f can be written as

$$\frac{f(x+) + f(x-)}{2} = a_0 + \sum_{n=1}^{+\infty} [a_n \cos(nx) + b_n \sin(nx)] \text{ for all } x \in \mathbb{R}.$$

$[f(x+) \text{ (resp. } f(x-)) \text{ denotes the right-limit (resp. the left-limit) of } f \text{ at } x]$.

4°) By choosing a particular x in the above expansion, determine the value of $\sum_{p=0}^{+\infty} \frac{(-1)^p}{2p+1}$.

5°) With the help of PARSEVAL's theorem, to be recalled, determine $\sum_{p=0}^{+\infty} \frac{1}{(2p+1)^2}$.

---

254. ★★ *The magic of* FOURIER *series (4)* (Easy)

Let $f : \mathbb{R} \to \mathbb{R}$ be a 2π-periodic function with

$$f(x) = |x| \text{ if } -\pi \leqslant x < \pi.$$

1°) Sketch the graph of f for x between -2π and 2π.

2°) Calculate the (real) FOURIER coefficients associated with f (denoted a_n and b_n).

3°) Give a clearly argued reason why f can be written as

$$f(x) = a_0 + \sum_{n=1}^{+\infty} [a_n \cos(nx) + b_n \sin(nx)] \text{ for all } x \in \mathbb{R};$$

$$|x| = \frac{\pi}{2} - \frac{4}{\pi} \sum_{p=0}^{+\infty} \frac{\cos[(2p+1)x]}{(2p+1)^2} \text{ for all } x \in [-\pi, +\pi].$$

72

4°) By choosing particular values of x in the above expansion, determine the value of $\sum_{p=0}^{+\infty} \frac{1}{(2p+1)^2}$, and also that of $\sum_{p=0}^{+\infty} \frac{1}{(2p+1)^2}(-1)^{\lfloor (p+1)/2 \rfloor}$, where $\lfloor x \rfloor$ denotes the integer part of x.

5°) With the help of PARSEVAL's theorem, to be recalled, determine $\sum_{p=0}^{+\infty} \frac{1}{(2p+1)^4}$.

---

255. ★★ *The magic of* FOURIER *series (5)* (Complex form)

Consider the following function: $f : \mathbb{R} \to \mathbb{R}$, periodic of period 2π, with

$$f(x) = e^x \text{ for } -\pi < x \leqslant \pi.$$

1°) For all $n = 0, \pm 1, \pm 2, ..., \pm n, ...$, calculate the complex FOURIER coefficients c_n, that is to say:

$$c_n = \frac{1}{2\pi} \int_{-\pi}^{\pi} f(x)e^{-inx}dx.$$

2°) Indicate briefly why

$$f(x) = \sum_{n=-\infty}^{+\infty} c_n e^{inx} \text{ for all } x \in \,]-\pi, \pi[.$$

3°) Deduce

$$e^x = \frac{\sinh \pi}{\pi} \sum_{n=-\infty}^{+\infty} (-1)^n \frac{1+in}{1+n^2} e^{inx} \text{ for all } x \in \,]-\pi, \pi[.$$

4°) Write the above expansion of e^x in the real form (that is to say, with sine and cosine functions of the real variable x).

---

256. ★★ *A discontinuous (infinite) sum of continuous functions*

Let us consider the following sequence $(u_n)_{n \geqslant 0}$ of functions defined on the whole of $\mathbb{R}$:

$$u_0(x) = x^2;$$
$$u_{n+1}(x) = \frac{1}{1+x^2}u_n(x).$$

1°) For a nonnegative integer N, we set $S_N(x) = \sum_{n=0}^{N} u_n(x)$.

(a) Show that

$$S_N(x) = 1 + x^2 - \frac{1}{(1+x^2)^N} \text{ for all } x \in \mathbb{R}.$$

73

(b) Deduce from the above that the series of functions with general term $(u_n)_{n \geqslant 0}$ is pointwise convergent and express $S(x) = \sum\limits_{n=0}^{+\infty} u_n(x)$ in a closed form.

2°) Give a simple reason why the convergence of the series of functions with general term $(u_n)_{n \geqslant 0}$ towards the function S is not uniform on $\mathbb{R}$.

---

257. ★★ *Change of variables: from Cartesian to polar coordinates*

Reminder on polar coordinates. For a point M, different from the origin, whose Cartesian coordinates are (x, y), the polar coordinates are the real numbers $(r, \theta) \in (0, +\infty) \times (-\pi, \pi]$ (uniquely) defined as follows:

r is the length of the vector $\overrightarrow{OM}$;

θ is the angle between the Ox axis and the vector $\overrightarrow{OM}$.

Here, the angles θ are measured in radians and positive in the counterclockwise sense.

Let $\Omega = (0, +\infty) \times (0, +\infty)$ and $U = (0, +\infty) \times (-\frac{\pi}{2}, \frac{\pi}{2})$. On Ω we can convert from Cartesian coordinates to polar coordinates, and vice versa, via the transformations

$$(x, y) \in \Omega \mapsto \left(r = \sqrt{x^2 + y^2}, \ \theta = \arctan\left(\frac{y}{x}\right)\right) \in U, \qquad (1\text{-}1)$$

$$(r, \theta) \in U \mapsto (x = r\cos(\theta), \ y = r\sin(\theta)) \in \Omega. \qquad (1\text{-}2)$$

The formulas above express a bijection and its inverse (both continuously differentiable) between Ω and U.

However, this is no longer true for larger sets Ω... Remember that the *arctan* function always returns an angle lying between $-\frac{\pi}{2}$ and $\frac{\pi}{2}$.

Let Ω_1 be the whole of $\mathbb{R}^2$ with the half-line of negative x cut out, that is to say

$$\begin{aligned} \Omega_1 &= \mathbb{R}^2 \backslash \{(x, y) : x \leqslant 0 \text{ and } y = 0\} \\ &= \{(x, y) : y \neq 0 \text{ or } x > 0\}. \end{aligned}$$

Let now $U_1 = (0, +\infty) \times (-\pi, \pi)$.

1°) Check that

$$F : (r, \theta) \in U_1 \mapsto (x = r\cos(\theta), \ y = r\sin(\theta)) \in \Omega_1$$

is a bijection from U_1 onto Ω_1.

2°) (a) Prove that the inverse bijection F^{-1} of F can be expressed, in a closed form, as follows:

$$F^{-1} : (x, y) \in \Omega_1 \mapsto \left(r = \sqrt{x^2 + y^2}, \ \theta = 2\arctan\left(\frac{y}{x + \sqrt{x^2 + y^2}}\right)\right) \in U_1. \quad (2)$$

74

(b) Check that both F and F^{-1} are continuously differentiable.

---

258. ★★ *Change of variables in a partial differential equation*

We look for twice continuously differentiable functions $u : (x, y) \in \Omega = (0, +\infty) \times (0, +\infty) \to \mathbb{R}$ satisfying

$$x\frac{\partial u}{\partial x} + y\frac{\partial u}{\partial y} = \sqrt{x^2 + y^2} \text{ for all } (x, y) \text{ in } \Omega. \tag{1}$$

1°) We pass from Cartesian coordinates (x, y) to polar coordinates (ρ, θ), so that $u(x, y)$ becomes $\widehat{u}(\rho, \theta)$.

Reformulate (1) in terms of $\widehat{u}$ and polar coordinates (ρ, θ).

2°) Deduce from this the general form of solutions of (1), written as functions u of Cartesian coordinates (x, y).

---

259. ★★ D'ALEMBERT's *formulation of solutions of a partial differential equation*

Consider the following partial differential equation, with boundary conditions:

$$(\mathcal{C}) \quad \begin{cases} \frac{\partial^2 u}{\partial t^2} = c^2 \frac{\partial^2 u}{\partial x^2} \\ u(x, 0) = f(x) \\ \frac{\partial u}{\partial t}(x, 0) = g(x), \end{cases}$$

where $c > 0$ is a given real number and both f and g are twice continuously differentiable functions of a real variable.

1°) Check that the function

$$u(x, t) = \frac{1}{2}\left[f(x + ct) + f(x - ct)\right] + \frac{1}{2c}\int_{x-ct}^{x+ct} g(\xi)\, d\xi \tag{1}$$

is a solution of $(\mathcal{C})$.

2°) Using (1), provide a solution of $(\mathcal{C})$ when $f = 0$ and $g(x) = \sin(x)$.

---

260. ★★ *An example of a uniqueness proof for a partial differential equation*

Given $L > 0$, real numbers a and b, and a function $f \in \mathcal{C}^2([0, L])$, one is interested in finding functions $(t, x) \in [0, +\infty) \times [0, L] \mapsto u(t, x)$ of class $\mathcal{C}^2$ satisfying:

$$\frac{\partial u}{\partial t}(t, x) - \frac{\partial^2 u}{\partial x^2}(t, x) \text{ for all } (t, x) \in [0, +\infty) \times [0, L],$$
$$u(t, 0) = a \text{ and } u(t, L) = b \text{ for all } t \in [0, +\infty), \text{ [boundary conditions]} \tag{1}$$
$$u(0, x) = f(x) \text{ for all } x \in [0, L], \text{ [initial condition]}.$$

We suppose that u_1 and u_2 are two solutions of (1) of class $\mathcal{C}^2$. We intend to prove that $u_1 = u_2$.

1°) Show that if $(t, x) \mapsto z(t, x)$ is a function of class $\mathcal{C}^2$, then

$$\frac{d}{dt}\left(\int_0^L z^2(t, x)\, dx\right) = 2\int_0^L z(t, x)\frac{\partial z}{\partial t}(t, x)\, dx.$$

2°) Let $v(t, x) = u_1(t, x) - u_2(t, x)$. Show that v is a solution to the same partial differential equation as above, but with simplified boundary conditions and initial condition.

3°) Let $t \geqslant 0 \mapsto E(t) = \frac{1}{2}\int_0^L v^2(t, x)\, dx$.

- Using the result obtained in the first question, prove that E is a decreasing function.

- Deduce that v is the null function.

---

261. ★★ *The telegraphist's partial differential equation*

Consider the following partial differential equation, called the "telegraphist's equation": find twice continuously differentiable functions $z : (t, x) \mapsto z(x, t)$ satisfying

$$\frac{\partial^2 z}{\partial x^2} - \frac{\partial^2 z}{\partial t^2} - 2\frac{\partial z}{\partial t} - z = 0. \qquad (1)$$

1°) We propose a change of functions:

$$u(x, t) = z(x, t)\, e^t.$$

(a) Express the partial derivatives $\frac{\partial z}{\partial t}, \frac{\partial^2 z}{\partial t^2}, \frac{\partial z}{\partial x}, \frac{\partial^2 z}{\partial x^2}$ in terms of partial derivatives of u.

(b) Deduce the partial differential equation equivalent to (1) where the unknown function is now u.

2°) Use the results above to provide a family of solutions of (1).

---

262. ★★ *Length of an ellipse*

Consider an ellipse $\mathcal{E}$ with equation, in Cartesian coordinates,

$$\frac{x^2}{a^2} + \frac{y^2}{b^2} = 1,$$

where $a \geqslant b > 0$.

1°) Calculate the area $\mathcal{A}(a, b)$ of the convex compact (elliptic) set C delimited by $\mathcal{E}$.

2°) We denote by $L(a, b)$ the length of the closed curve $\mathcal{E}$.

(a) Provide an expression for $L(a, b)$ in terms of an integral or a sum of a series.

(b) KEPLER proposed to approximate $L(a, b)$ with the help of the geometric and arithmetic means of a and b:

$$L(a, b) \approx 2\pi\sqrt{ab} \text{ and } L(a, b) \approx 2\pi\frac{a+b}{2} = \pi(a+b). \tag{1}$$

EULER proposed another approximation of $L(a, b)$:

$$L(a, b) \approx 2\pi\sqrt{\frac{a^2 + b^2}{2}} = \pi\sqrt{2(a^2 + b^2)}. \tag{2}$$

Compare these two approximations with the exact value of $L(a, b)$.

---

263. ★★ *Movement of the center of an ellipse stuck to the axes of the first quadrant*

Consider an ellipse authorized to move in the positive (or first) quadrant of the plane while remaining tangent to the coordinate axes. Question: what is the path followed by its center?

Put in more mathematical terms: if $\mathcal{E}$ is an ellipse with axial parameters $a \geqslant b > 0$, contained in the positive quadrant $\{(x, y) : x \geqslant 0 \text{ and } y \geqslant 0\}$, tangent to the horizontal axis Ox as well as to the vertical axis Oy; then, what is the locus of the center Ω of this ellipse?

---

264. ★★ *A mind-blowing characterization of the sine function*

Prove that the sine function is the only 2π-periodic function $f \in \mathcal{C}^\infty(\mathbb{R}, \mathbb{R})$ satisfying:

$$f'(0) = 1; \tag{1}$$
$$\left|f^{(n)}(x)\right| \leqslant 1 \text{ for all } x \in \mathbb{R} \text{ and all integers } n \geqslant 0. \tag{2}$$

---

265. ★★ *A characterization of power functions among convex functions*

Let $f : [0, +\infty) \to \mathbb{R}$ be a convex function that we assume (right-) continuous at 0. We intend to determine, among such functions f, those which satisfy furthermore:

$$f(x)f(y) \leqslant f(xy) \text{ for all } x \geqslant 0 \text{ and } y \geqslant 0, \tag{1}$$
$$f(1) = 1. \tag{2}$$

Obviously, the constant function $f = 1$ satisfies (1)–(2).

Prove that the non-constant functions f satisfying the requirements above are the power functions $x \geqslant 0 \mapsto f(x) = x^p$, with $p \geqslant 1$.

"In mathematics, the art of proposing a question must be held of higher value than solving it"

G. CANTOR (1845–1918)

266. ★★★ *The cardinalities of point-sets in space vs their planar projections*

In the three-dimensional space $\mathbb{R}^3$, with coordinate system $(Oxyz)$, consider a finite set of points $\mathcal{F}$. We denote by $\mathcal{F}_x, \mathcal{F}_y, \mathcal{F}_z$ the sets of orthogonal projections of $\mathcal{F}$ on the planes $(Oyz), (Oxz), (Oxy)$, respectively.

Prove that

$$(card\ \mathcal{F})^2 \leqslant (card\ \mathcal{F}_x) \times (card\ \mathcal{F}_y) \times (card\ \mathcal{F}_z), \tag{1}$$

where $card\ S$ denotes the number of elements in the finite set S.

---

267. ★★★ *"Numerically equal" or "mathematically equal"? (1)*

Consider the two following real numbers:

$$A = \sqrt{5} + \sqrt{22 + 2\sqrt{5}};$$

$$B = \sqrt{11 + 2\sqrt{29}} + \sqrt{2\sqrt{55 - 10\sqrt{29}} + 16 - 2\sqrt{29}}.$$

Numerical computations show that A and B are "equal", in the sense that their decimal expansions are the same, for dozens and dozens of digits...

But, are A and B equal?

---

268. ★★★ *"Numerically equal" or "mathematically equal"? (2)*

Consider the following "large" integers:

$$S = (3987)^{12}\ ;\ T = (4365)^{12}$$
$$U = (4472)^{12}.$$

1°) Compute, using the PYTHON programming language for example, $S + T$ and U.

2°) Calculate the relative difference $\varepsilon_r = \frac{(S+T)-U}{S+T+U}$.

---

269. ★★★ *A peculiar property of the integer* 26

The integer 26 is trapped between a squared integer ($25 = 5^2$) and a cubed integer ($27 = 3^3$).

Prove that this is the only case among positive integers.

---

270. ★★★ *Raising irrational (or rational) numbers to irrational (or rational) powers*

The real number $\sqrt{2}$ is irrational. Is $\left(\sqrt{2}\right)^{\sqrt{2}}$ rational or irrational? We do not know...; at least, we cannot answer the question with elementary methods. Nevertheless, we can use the numbers $\sqrt{2}$ and $\sqrt{2}$ raised to powers of $\sqrt{2}$ to answer the following questions:

- Can an irrational number raised to an irrational power be rational?

- Can an irrational number raised to an irrational power be irrational ?

---

271. ★★★ Let $u_0, u_1, ..., u_n, ...$ be a sequence taking values in the set $\mathbb{N}$ (the nonnegative integers). Suppose that

$$u_{n+1} > u_{u_n} \text{ for all } n \in \mathbb{N}. \tag{1}$$

What can be said about such a sequence (u_n)?

---

272. ★★★ Let $(u_n)_{n \geqslant 1}$ be a sequence of real numbers converging to l. Consider the sequence $(v_n)_{n \geqslant 1}$ defined from $(u_n)_{n \geqslant 1}$ as follows:

$$\text{For } n \geqslant 1, \ v_n = \frac{u_1 + 2u_2 + ... + nu_n}{1 + 2 + ... + n}.$$

(v_n is a particular convex combination of the n terms $u_1, u_2, ..., u_n$, with highest weights on the latest terms.)

Show that the sequence $(v_n)_{n \geqslant 1}$ converges to the same limit l.

---

273. ★★★ *A power series built up from numbers of partitions*

Let D_n denote the number of partitions of $\{1, 2, ..., n\}$, with $D_0 = 1$ by convention.

1°) Determine D_1, D_2, D_3.

2°) Prove the following recursive formula:

$$\text{For all } n \geqslant 0, \ D_{n+1} = \sum_{k=0}^{n} \binom{n}{k} D_k. \tag{1}$$

3°) Prove that the power series with general term $\frac{D_k}{k!} x^k$ has a convergence radius $R \geqslant 1$, and calculate the sum $f(x) = \sum_{k=0}^{+\infty} \frac{D_k}{k!} x^k$ for $|x| < R$.

79

274. ★★★ KRASNOSELSKI's *iteration scheme* (1955)

Let $f : [a, b] \to [a, b]$ satisfy a 1-LIPSCHITZ condition on $[a, b]$, that is to say:

$$|f(x) - f(y)| \leqslant |x - y| \text{ for all } x, y \text{ in } [a, b].$$

Define an iteration scheme as follows:

$$\begin{cases} x_0 \in [a, b] \text{ (initial point)}, \\ x_{n+1} = \frac{1}{2}[x_n + f(x_n)] \text{ for nonnegative integers } n. \end{cases}$$

1°) Prove that (x_n) is a convergent sequence in $[a, b]$ and that its limit is a fixed point of f.

2°) Show with the help of a counterexample that the result does not hold true if we just assume that $f : [a, b] \to [a, b]$ is continuous.

---

275. ★★★ FEJER *sequences*

Let $f : \mathbb{R}^n \to \mathbb{R}^n$ be a continuous mapping and let S be a nonempty subset of $\mathbb{R}^n$. We say that f is S-FEJER when the following two conditions are fulfilled:

$$\begin{aligned} f(y) &= y \text{ for all } y \in S; \\ \|f(x) - y\| &< \|x - y\| \text{ for all } x \notin S \text{ and } y \in S. \end{aligned}$$

A simple example of such a situation is: $S = \{0\}$ in $\mathbb{R}$ and $f : \mathbb{R} \to \mathbb{R}$ defined by $f(x) = \frac{x}{2}$.

Define a sequence $(x_k)_k$ of $\mathbb{R}^n$ as follows:

$$\begin{cases} x_0 \text{ arbitrary in } \mathbb{R}^n; \\ x_{k+1} = f(x_k) \text{ for all nonnegative integers } k. \end{cases}$$

Prove that (x_k) is convergent and that its limit point belongs to S.

---

276. ★★★ *A more difficult functional equation (1)*

We look for twice differentiable functions $f : \mathbb{R} \to \mathbb{R}$ satisfying:

$$f(x + y)f(x - y) = [f(x)]^2 + [f(y)]^2 - 1 \text{ for all } x, y \text{ in } \mathbb{R}. \tag{1}$$

1°) Give simple examples of functions f satisfying (1).

2°) Show that any f satisfying (1) is necessarily a solution of the following linear second-order differential equation with constant coefficients:

$$f'' \pm k^2 f = 0. \tag{2}$$

3°) Deduce from the above all the solutions of the functional equation (1).

277. ★★★ *A more difficult functional equation (2)*

We look for twice differentiable functions $f : \mathbb{R} \to \mathbb{R}$ satisfying:

$$f(x)f(y) = f(\sqrt{x^2 + y^2}) \text{ for all } x, y \text{ in } \mathbb{R}, \qquad (1\text{-}1)$$
$$f(x) \to 0 \text{ as } |x| \to +\infty. \qquad (1\text{-}2)$$

1°) Give an example of a non-null function f satisfying (1-1).

2°) Let f satisfy (1-1). Show that either $f(0) = 0$ or $f(1) = 1$.

3°) Let f satisfy (1-1) and $f(0) = 0$. Show that f is the null function.

4°) Let f satisfy (1-1) and $f(0) = 1$.

(a) Prove that f solves the following linear first-order differential equation:

$$f'(x) = xf''(0)f(x), \ x \in \mathbb{R}. \qquad (2)$$

(b) Deduce from the above all the functions solving (1-1) and (1-2).

---

278. ★★★ *A "mixture" of two easy linear second-order differential equations*

Let b be a given real number.

1°) First solve the following two easy linear CAUCHY problems:

$$(\mathcal{C}_1) \quad y'' + y = 0; \ y(0) = -1; \ y'(0) = b.$$
$$(\mathcal{C}_2) \quad y'' - y = 0; \ y(0) = -1; \ y'(0) = b.$$

2°) Now solve the following nonlinear CAUCHY problem, a kind of "mixture" of the two previous ones:

$$(\mathcal{C}_3) \quad y'' + |y| = 0; \ y(0) = -1; \ y'(0) = b.$$

---

279. ★★★ *Qualitative analysis of a "highly" nonlinear differential equation*

Consider the following first-order nonlinear differential equation :

$$(ED) \quad y' + y^3 = \sin(t).$$

It turns out that (ED) has a unique solution which is defined on the whole of $\mathbb{R}$; this solution is periodic.

Prove the following:

(a) When $t \to +\infty$, any solution of (ED) ends up by entering the interval $(-1, 1)$ and staying there, *i.e.* there exists a τ such that

$$|y(t)| < 1 \text{ for } t > \tau.$$

81

(b) When $t \to +\infty$, two solutions y_1 and y_2 of (ED) become closer and closer:

$$y_1(t) - y_2(t) \to 0 \text{ as } t \to +\infty.$$

(c) When $t \to -\infty$, all the solutions y of (ED), except the periodic one, end by "blowing up". For such a solution y, there exists a τ such that

$$|y(t)| \to +\infty \text{ when } t > \tau \to \tau.$$

---

280. ★★★ *A "non-standard" first-order differential equation*

Standard first-order differential equations are written in the form $y'(t) = ...$ We consider here a non-standard one:

$$(ED) \quad \varphi[y'(t)] + y(t) = \cos(t),$$

where $\varphi(x) = \min(1, |x|)$.

Show that (ED) has no solution on the interval $[-\frac{\pi}{2}, \frac{\pi}{2}]$.

---

281. ★★★ *Perturbing polynomials having only real roots*

Let P be a non-null real polynomial having only real roots. We perturb it by adding the polynomial rP',

$$P_r = P + rP',$$

where r is a real number.

Show that, like P, P_r has only real roots.

---

282. ★★★ *The intermediate value property vs continuity for functions*

A continuous function $f : \mathbb{R} \to \mathbb{R}$ is known to satisfy the intermediate value property, that is: the image by f of any interval I is an interval. But, a function satisfying the intermediate value property is not necessarily continuous, an example is:

$$\varphi(x) = \begin{cases} \sin\left(\frac{1}{x}\right) & \text{if } x \neq 0, \\ 0 & \text{if } x = 0. \end{cases} \tag{1}$$

So, we would like to answer the following question: what exactly is missing for a function satisfying the intermediate value property to be continuous?

Given $f : \mathbb{R} \to \mathbb{R}$, prove that f is continuous if and only if:

$$\begin{cases} \quad (a) \ f \text{ satisfies the intermediate value property;} \\ (b) \text{ For all } r \in \mathbb{R}, \text{ the set } \{x \in \mathbb{R} : f(x) = r\} \text{ is closed.} \end{cases} \tag{2}$$

---

283. ★★★ *Primitive functions (or antiderivatives) of some non-continuous functions*

Let $f : \mathbb{R} \to \mathbb{R}$ be continuous, not identically equal to 0 and periodic. We denote by $T > 0$ a period of f.

The function g defined for $x \neq 0$ by $g(x) = f(\frac{1}{x})$ is continuous (at any $x \neq 0$), but there is no way to extend it at 0 so that the extended function is continuous at 0 (this is easy to check).

We therefore ask the question of how to extend g at 0 so that the extended function has an antiderivative (or is primitivable) on $\mathbb{R}$.

1°) (a) We extend g at 0 by setting $g(0) = \frac{1}{T} \int_0^T f(t)dt$. Prove that the function g, so extended to the whole of $\mathbb{R}$, has an antiderivative on $\mathbb{R}$.

(b) Show that g, extended at 0 by another value than $\frac{1}{T} \int_0^T f(t)dt$, has no antiderivative on $\mathbb{R}$.

2°) Examples. How can one extend the following functions $x \neq 0 \mapsto g(x) = f(\frac{1}{x})$ at 0 such that the extended function has an antiderivative on $\mathbb{R}$?

$g(x) = \sin\left(\frac{1}{x}\right)$; $g(x) = \left|\sin\left(\frac{1}{x}\right)\right|$; $g(x) = \sin^2\left(\frac{1}{x}\right)$.

---

284. ★★★ *A hemi-differential equation*

Let $f : \mathbb{R} \to \mathbb{R}$ be a function having (finite) right-derivatives and left-derivatives at all points. Such one-sided derivatives are denoted by f'_+ and f'_-, respectively.

1°) Show that f is necessarily continuous.

2°) We suppose that

$$f'_+(x) = 2f'_-(x) \text{ for all } x \in \mathbb{R}. \tag{1}$$

Prove that f is a constant function.

---

285. ★★★ *Limit of the integral of a product vs the product of integrals*

Let f and $g : \mathbb{R} \to \mathbb{R}$ be two continuous 1-periodic functions. Then prove that:

$$\int_0^1 f(x)g(nx) \, dx \to \left[\int_0^1 f(x) \, dx\right] \times \left[\int_0^1 g(x) \, dx\right] \text{ when } n \to +\infty.$$

---

286. ★★★ *A generalized discriminant*

Let $P(x)$ be a polynomial function of degree $n \geqslant 1$, with real coefficients, having only real roots. For such a polynomial, we define:

$$\Delta_n(x) = (n-1)\left[P'(x)\right]^2 - nP(x)P''(x).$$

1°) What is $\Delta_1(x)$? $\Delta_2(x)$?

2°) Show that
$$\Delta_n(x) \geqslant 0 \text{ for all } x \in \mathbb{R}.$$

---

287. ★★★ *On the mean value theorem*

1°) Find all the differentiable functions $f : \mathbb{R} \to \mathbb{R}$ satisfying the following mean value property:

$$\frac{f(x) - f(y)}{x - y} = f'\left(\frac{x + y}{2}\right) \text{ for all } x \neq y \text{ in } \mathbb{R}. \tag{1}$$

2°) Choose now two nonnegative real numbers α and β with sum 1, but different from $1/2$. Then find all the differentiable functions $f : \mathbb{R} \to \mathbb{R}$ satisfying the following mean value property:

$$\frac{f(x) - f(y)}{x - y} = f'(\alpha x + \beta y) \text{ for all } x \neq y \text{ in } \mathbb{R}. \tag{2}$$

[We do not impose here that $c = \alpha x + \beta y$ is the only point in (a, b) for which (2) holds true.]

---

288. ★★★ C^2 *functions are diff-convex*

Let $f : \mathbb{R}^n \to \mathbb{R}$ be a function of class C^2.

1°) Prove that one can find convex functions $g, h : \mathbb{R}^n \to \mathbb{R}$ such that $f = g - h$.

2°) Show that g, for example, can be chosen of class C^2 and h of class C^∞.

---

289. ★★★ *An integral equation on continuous functions*

Let $f : \mathbb{R} \to \mathbb{R}$ be a continuous function and consider the following two conditions:

$$f(x) = \frac{1}{2} \int_{x-1}^{x+1} f(t) \, dt \text{ for all } x \in \mathbb{R}; \tag{1}$$

$$\lim_{x \to +\infty} f(x) = \lim_{x \to -\infty} f(x) = 0. \tag{2}$$

1°) Give an example of a non-null function f satisfying (1).

2°) Show that there is no polynomial function f of degree $n \geqslant 2$ satisfying (1).

3°) Prove that a continuous $f : \mathbb{R} \to \mathbb{R}$ satisfying both (1) and (2) is necessarily the null function.

84

290. ★★★ *A way to ensure that* $\lim_{x \to +\infty} f(x) = 0$ *via limits at discrete points*

Let $f : (1, +\infty) \to \mathbb{R}$ be a continuous function. We consider a sequence (α_n) of real numbers satisfying:

$$\begin{cases} 1 \leqslant \alpha_1 < \alpha_2 < \ ... \ < \alpha_n \to +\infty; \\ \lim_{n \to +\infty} \frac{\alpha_{n+1}}{\alpha_n} = 1. \end{cases} \tag{1}$$

We suppose:

$$\text{for all } x > 1, \ \lim_{n \to +\infty} f(\alpha_n x) = 0. \tag{2}$$

Prove that $\lim_{x \to +\infty} f(x) = 0$.

---

291. ★★★ Let $f : [a, b] \to (0, +\infty)$ be a continuous function. We suppose that the parallel lines to the Oy axis, whose abscissas are the $n + 1$ points

$$x_0 = a < x_1 < \ ... \ < x_n = b$$

divide the area $\int_a^b f(x) \, dx$ into n equal areas.

Prove that the mean value $M_n = \frac{1}{n} \sum_{i=1}^{n} f(x_i)$ of the function values at these specific points has a limit as $n \to +\infty$.

---

292. ★★★ *Inequalities between the "energies" of a function and of its derivative*

Let $f : [0, 1] \to \mathbb{R}$ be continuously differentiable on $[0, 1]$ and satisfy the end-point constraints $f(0) = f(1) = 0$. We intend to prove that

$$\int_0^1 [f(x)]^2 \, dx \leqslant \int_0^1 [f'(x)]^2 \, dx. \tag{1}$$

Here is a proposed approach.

1°) Let $y : [0, 1] \to \mathbb{R}$ be continuously differentiable on $[0, 1]$. Show that

$$\begin{cases} \int_0^1 2yff' \, dx = -\int_0^1 f^2 y' \, dx \\ \text{(integrate by parts and use the assumption } f(0) = f(1) = 0); \\ \int_0^1 (f' - yf)^2 \, dx = \int_0^1 [(f')^2 + (y' + y^2)f^2] \, dx. \end{cases} \tag{2}$$

2°) Suppose now that y solves the following differential equation

$$y' + y^2 = -1.$$

An example of such a y is $y(x) = -\tan x$.

Conclude with the relations in (2).

---

293. ★★★ *Images of slopes of vector-valued functions vs those of their derivatives*

Let $f : \mathbb{R} \to \mathbb{R}^2$ be the function defined by $f(t) = (\cos t, \sin t)$. It is clear that its derivative (vector) $f'(t)$ is $(-\sin t, \cos t)$. Hence the image $f'([0, 2\pi])$ of $[0, 2\pi]$ by the derivative function f' is the unit circle in $\mathbb{R}^2$. Consider now the (vector made of) difference quotients (or slopes) of f :

$$q = \frac{f(s) - f(t)}{s - t} \text{ with } 0 \leqslant t < s \leqslant 2\pi.$$

1°) By choosing $u = \frac{s+t}{2}$ and $v = \frac{s-t}{2}$, verify that

$$q = \frac{\sin v}{v} \left(\cos \left(u + \frac{\pi}{2} \right), \sin \left(u + \frac{\pi}{2} \right) \right), \tag{1}$$

so that the (usual) Euclidean length of q is strictly less than 1.

2°) Determine and draw the set

$$Q = \left\{ q = \frac{f(s) - f(t)}{s - t} \text{ with } 0 \leqslant t < s \leqslant 2\pi \right\}.$$

Check that the closed convex hull of Q and that of $f'([0, 2\pi])$ are the same, namely the closed unit ball of $\mathbb{R}^2$.

---

294. ★★★ *When the C^∞-differentiability of powers of f implies that of f*

Let $f : \mathbb{R} \to \mathbb{R}$ be given.

1°) Suppose there exists a positive integer n_0 such that f^n is of class C^∞ for all $n \geqslant n_0$. Prove that f itself is of class C^∞.

2°) Suppose that just f^2 and f^3 are of class C^∞. Prove that f itself is of class C^∞.

---

295. ★★★ *Finite groups of matrices whose traces sum up to 0*

Let $G = \{M_1, M_2, ..., M_d\}$ be a finite set of matrices in $\mathcal{M}_n(\mathbb{C})$, which form a group under matrix multiplication. We suppose that

$$\sum_{i=1}^{d} \text{trace}(M_i) = 0.$$

Show that

$$\sum_{i=1}^{d} M_i = 0.$$

---

296. ★★★ *A topological characterization of diagonalizability*

Let $M \in \mathcal{M}_n(\mathbb{C})$ and let $\mathcal{C}(M)$ denote its "class of similarity", that is to say:

$$\mathcal{C}(M) = \left\{ P^{-1}MP \mid P \text{ is an invertible matrix} \right\}.$$

Prove that M is diagonalizable if and only if $\mathcal{C}(M)$ is a closed set (in $\mathcal{M}_n(\mathbb{C})$).

---

297. ★★★ *Functions of invariants of a matrix*

For $M \in \mathcal{M}_n(\mathbb{C})$, let P_M denote its characteristic polynomial:

$$\begin{aligned} P_M(\lambda) &= \det(M - \lambda I_n) \\ &= (-1)^n \left[\lambda^n - c_1(M)\lambda^{n-1} + c_2(M)\lambda^{n-2} - \ldots + (-1)^n c_n(M) \right]. \end{aligned}$$

The coefficients $c_k(M)$ are called *the invariants of the matrix M*. For example, $c_1(M)$ is the trace of M (therefore the sum of all eigenvalues λ_k of M), $c_n(M)$ is the determinant of M (therefore the product of all eigenvalues λ_k of M).

1°) Check that $c_k(M) = c_k(P^{-1}MP)$ for every invertible P in $\mathcal{M}_n(\mathbb{C})$.

2°) Suppose that we have a continuous function $f : \mathcal{M}_n(\mathbb{C}) \to \mathbb{C}$ satisfying:

$$f(P^{-1}MP) = f(M) \tag{1}$$

for every M in $\mathcal{M}_n(\mathbb{C})$ and every invertible P in $\mathcal{M}_n(\mathbb{C})$.

Prove that there exists a unique function $F_f : \mathbb{C}^n \to \mathbb{C}$ such that:

$$f(M) = F_f(c_1(M), c_2(M), \ldots, c_n(M)) \text{ for all } M \text{ in } \mathcal{M}_n(\mathbb{C}). \tag{2}$$

---

298. ★★★ *Convex hull of the set of nilpotent matrices*

Consider the set $\mathcal{N}_n$ of nilpotent matrices of $\mathcal{M}_n(\mathbb{R})$.

1°) Show that $\mathcal{N}_n$ has an empty interior, is closed and arcwise-connected (even star-shaped with respect to 0).

2°) Prove that the convex hull of $\mathcal{N}_n$ is the vector space of matrices whose trace is 0.

---

299. ★★★ *Finding the determinant of a matrix by blocks (2)*

Consider a family $(A_{i,j})_{1 \leqslant i,j \leqslant n}$ of matrices of $\mathcal{M}_m(\mathbb{R})$ such that any two of them commute. Let A be the matrix in $\mathcal{M}_{m \times n}(\mathbb{R})$ built up from blocks $A_{i,j}$ as follows:

$$A = \begin{bmatrix} A_{1,1} & \ldots & \ldots & A_{1,n} \\ A_{2,1} & \ldots & \ldots & A_{2,n} \\ \ldots & \ldots & \ldots & \ldots \\ A_{n,1} & \ldots & \ldots & A_{n,n} \end{bmatrix}.$$

Prove that

$$\det A = \det \left(\sum_\sigma \epsilon(\sigma) A_{\sigma(1),1} \dots A_{\sigma(n),n} \right), \tag{1}$$

where the summation is taken over all the permutations σ of $\{1, 2, ..., n\}$ and $\epsilon(\sigma)$ stands for the signature of σ.

Formula (1) hence generalizes the (original) formula giving the determinant of

$$A = \begin{bmatrix} a_{1,1} & \dots & \dots & a_{1,n} \\ a_{2,1} & \dots & \dots & a_{2,n} \\ \dots & \dots & \dots & \dots \\ a_{n,1} & \dots & \dots & a_{n,n} \end{bmatrix},$$

where the $a_{i,j}$ are just real numbers.

300. ★★★ *Behavior of the eigenvalues and eigenvectors of matrices depending on parameters*

1°) Example 1 (by J.W. GIVENS). Consider the following 2×2 symmetric real matrix

$$M(\theta) = \begin{bmatrix} 1 + \theta \cos\left(\frac{2}{\theta}\right) & -\theta \sin\left(\frac{2}{\theta}\right) \\ -\theta \sin\left(\frac{2}{\theta}\right) & 1 - \theta \cos\left(\frac{2}{\theta}\right) \end{bmatrix} \text{ if } \theta \neq 0; \ M(0) = \begin{bmatrix} 1 & 0 \\ 0 & 1 \end{bmatrix}.$$

Clearly, $M(\theta) \to M(0)$ as $\theta \to 0$.

What are the eigenvalues and the associated eigenspaces of $M(\theta)$ for $\theta \neq 0$? What happens when $\theta \to 0$?

2°) Example 2 (by F. RELLICH). Consider the following 2×2 symmetric real matrix

$$N(\theta) = e^{-1/\theta^2} \begin{bmatrix} \cos\left(\frac{2}{\theta}\right) & \sin\left(\frac{2}{\theta}\right) \\ \sin\left(\frac{2}{\theta}\right) & -\cos\left(\frac{2}{\theta}\right) \end{bmatrix} \text{ if } \theta \neq 0; \ N(0) = \begin{bmatrix} 0 & 0 \\ 0 & 0 \end{bmatrix}.$$

Clearly, $N(\theta) \to N(0)$ when $\theta \to 0$.

What are the eigenvalues and the associated eigenspaces of $N(\theta)$ for $\theta \neq 0$? What happens when $\theta \to 0$?

301. ★★★ *Possible diagonalization of antidiagonal matrices*

The matrix

$$M = \begin{bmatrix} 0 & 0 & \dots & 0 & a_n \\ 0 & \dots & 0 & a_{n-1} & 0 \\ \dots & 0 & \dots & 0 & \dots \\ 0 & a_2 & 0 & \dots & 0 \\ a_1 & 0 & \dots & 0 & 0 \end{bmatrix} \in \mathcal{M}_n(\mathbb{C}),$$

88

where $a_1, a_2, ..., a_n \in \mathbb{C}$, looks similar to a diagonal matrix since it is antidiagonal. But it is not always diagonalizable.

What are the conditions, expressed in terms of the "antidiagonal" coefficients $a_1, a_2, ..., a_n$, necessary and sufficient for M to be diagonalizable?

---

302. ★★★ *Matrices as linear combinations of a small number of orthogonal matrices*

1°) Check that the linear space generated by the set of $n \times n$ orthogonal matrices is the whole space $\mathcal{M}_n(\mathbb{R})$.

2°) Let now $\pi(n)$ denote the least integer p such that any matrix in $\mathcal{M}_n(\mathbb{R})$ can be written as a linear combination of at most p orthogonal matrices.

(a) What is $\pi(1)$? $\pi(2)$?

(b) Show that $\pi(3) = 3$.

(c) Prove that:
$$\pi(n) = 3 \text{ or } 4 \text{ for all } n \geqslant 4.$$

Consequently, any matrix in $\mathcal{M}_n(\mathbb{R})$ can be written as a linear combination of 4 orthogonal matrices.

---

303. ★★★ $\mathbb{R}$ *vs* $\mathcal{S}_n(\mathbb{R})$: *similarities and dissimilarities*

Let $\mathcal{S}_n(\mathbb{R})$ be equipped with the ordering $\succcurlyeq$ defined as follows: $A \succcurlyeq B$ if and only if $A - B$ is positive semidefinite ($\succcurlyeq 0$ in short).

1°) Consider a sequence (A_k) in $\mathcal{S}_n(\mathbb{R})$ which is increasing and bounded from above (both properties are expressed with the ordering $\succcurlyeq$). Is this sequence convergent?

2°) Let S be a set in $\mathcal{S}_n(\mathbb{R})$ which is bounded from below (that is to say: there exists an $M \in \mathcal{S}_n(\mathbb{R})$ such that $A \succcurlyeq M$ for all A in S). Does there exist a greatest lower bound (or infimum) for S?

3°) Is the function $A \mapsto A^2$, defined on the set of positive semidefinite matrices, increasing?

4°) Is the function $A \mapsto A^{-1}$, defined on the set of positive definite matrices, decreasing?

---

304. ★★★ *A modified* CAUCHY–SCHWARZ *inequality*

1°) The discrete case. Let $\mathbb{R}^n$ be equipped with the usual inner-product and the associated Euclidean norm. Let $x = (x_1, x_2, \dots, x_n)$ and $(y_1, y_2, \dots, y_n)$ be two vectors in $\mathbb{R}^n$.

Prove that
$$|x^T y| \leqslant \sqrt{\left(\sum_{i=1}^{n} \max(x_i^2, y_i^2) \right) \times \left(\sum_{i=1}^{n} \min(x_i^2, y_i^2) \right)} \qquad (1)$$

and

$$\|x\|^2 \times \|y\|^2 = \left(\sum_{i=1}^{n} \max(x_i^2, y_i^2) \right) \times \left(\sum_{i=1}^{n} \min(x_i^2, y_i^2) \right) \tag{2}$$

$$+ \left(\sum_{\{i \,:\, x_i^2 > y_i^2\}} (x_i^2 - y_i^2) \right) \times \left(\sum_{\{i \,:\, x_i^2 < y_i^2\}} (y_i^2 - x_i^2) \right).$$

2°) The continuous case. Let f and g be two functions in $\mathcal{C}([a, b], \mathbb{R})$. Prove that

$$\left| \int_a^b f(x)g(x) \, dx \right| \leqslant \sqrt{ \left(\int_a^b \max(f^2, g^2)(x) \, dx \right) \times \left(\int_a^b \min(f^2, g^2)(x) \, dx \right) } \tag{3}$$

and

$$\left(\int_a^b f^2(x) \, dx \right) \times \left(\int_a^b g^2(x) \, dx \right) =$$

$$\left(\int_a^b \max(f^2, g^2)(x) \, dx \right) \times \left(\int_a^b \min(f^2, g^2)(x) \, dx \right)$$

$$+ \left(\int_a^b \left(f^2 - g^2 \right)^+ (x) \, dx \right) \times \left(\int_a^b \left(g^2 - f^2 \right)^+ (x) \, dx \right). \tag{4}$$

---

305. ★★★ *A generalized* CAUCHY–SCHWARZ *inequality*

Let A and M be two symmetric real (n, n) matrices satisfying:

$$\left| x^T M x \right| \leqslant x^T A x \text{ for all } x \in \mathbb{R}^n. \tag{1}$$

Prove that

$$(a^T M b)^2 \leqslant (a^T A a) \times (b^T A b) \text{ for all } a \text{ and } b \text{ in } \mathbb{R}^n. \tag{2}$$

---

306. ★★★ *Regions of nonnegativity of quadratic forms associated with A and A^{-1}*

Let $A \in \mathcal{M}_n(\mathbb{R})$ be symmetric and invertible.

1°) What are the conditions on A which imply

$$\left\{ x \in \mathbb{R}^n \mid x^T A x \geqslant 0 \right\} \subset \left\{ x \in \mathbb{R}^n \mid x^T A^{-1} x \geqslant 0 \right\}? \tag{1}$$

2°) Consequence: what are the conditions on A which imply

$$\left\{ x \in \mathbb{R}^n \mid x^T A x \geqslant 0 \right\} = \left\{ x \in \mathbb{R}^n \mid x^T A^{-1} x \geqslant 0 \right\}? \tag{2}$$

---

307. ★★★ *Any matrix can be written as the product of two symmetric matrices*

Let $\mathcal{S}_n(\mathbb{R})$ and $\mathcal{AS}_n(\mathbb{R})$ denote respectively the set of $n \times n$ real symmetric matrices and the set of $n \times n$ real antisymmetric matrices. Given $A \in \mathcal{M}_n(\mathbb{R})$, we consider the linear transformation $\varphi : \mathcal{S}_n(\mathbb{R}) \to \mathcal{AS}_n(\mathbb{R})$ defined by:

$$\varphi(S) = AS - SA^T.$$

1°) Proving that A is similar to its transpose A^T.

- Show that the dimension of the kernel of φ is at least n.

- Use this property to prove that there exists a nonsingular symmetric matrix S such that

$$S^{-1}AS = A^T. \tag{1}$$

Hence any matrix is similar to its transpose.

2°) Use the property proved in the question above to show that any $A \in \mathcal{M}_n(\mathbb{R})$ can be written as the product of two symmetric matrices:

$$A = S_1 S_2, \text{ with } S_1 \text{ and } S_2 \text{ in } \mathcal{S}_n(\mathbb{R}). \tag{2}$$

---

308. ★★★ *Determinant and trace inequalities for 2×2 positive definite matrices*

For two real symmetric positive definite 2×2 matrices A and B, let

$$\theta(A, B) = \det(A + B) - \det(A) - \det(B).$$

1°) Show that

$$\theta(A, B) = (\operatorname{tr} A) \times (\operatorname{tr} B) - \operatorname{tr}(AB)$$
$$\text{and}$$
$$\theta(A, B) > 0.$$

2°) (a) Prove the equivalence of the following three statements:

$$\theta(A, B) \times (\operatorname{tr} A) \times (\operatorname{tr} B) \geqslant [\theta(A, B)]^2 + [\operatorname{tr} A]^2 \times \det B; \tag{1}$$
$$\theta(A, B) \times [\operatorname{tr} B + \det A] \times \operatorname{tr} A > [\theta(A, B)]^2 + [\operatorname{tr} A]^2 \times \det B; \tag{2}$$
$$\left[\frac{(\operatorname{tr} A)^2}{2}\right] \times \left[\operatorname{tr}(B^2) - \frac{(\operatorname{tr} B)^2}{2}\right] \geqslant \left[\operatorname{tr}(AB) - \frac{(\operatorname{tr} A) \times \operatorname{tr}(B)}{2}\right]^2. \tag{3}$$

(b) Prove the inequality (3).

---

309. ★★★ A "determinantal" characterization of $AB = 0$ for two symmetric matrices A and B

Let A and B be two real symmetric $n \times n$ matrices.

Prove the equivalence of the following three statements:

$$AB = 0; \tag{1}$$

$$\det(I_n + xA + xB) = \det(I_n + xA)\det(I_n + xB) \text{ for all } x \in \mathbb{R}; \tag{2}$$

$$\det(I_n + xA + yB) = \det(I_n + xA)\det(I_n + yB) \text{ for all } x, y \in \mathbb{R}. \tag{3}$$

---

310. ★★★ A characterization of commutativity via the exponentials of matrices

Let A and B be two matrices in $\mathcal{M}_n(\mathbb{C})$.

1°) Suppose that $AB = BA$. Show that

$$\exp(mA + nB) = (\exp A)^m \times (\exp B)^n \text{ for all } m, n \text{ in } \mathbb{Z}. \tag{1}$$

2°) Converse. Prove that the relation (1) implies that A and B commute.

---

311. ★★★ A product of positive definite matrices, if symmetric, is positive definite

Let S_1 and S_2 be two symmetric positive definite matrices. We suppose that $S_1 S_2$ is symmetric. We want to show that $S_1 S_2$ is positive definite. For that purpose, let us define, for $0 \leqslant t \leqslant 1$,

$$
\begin{aligned}
M(t) &= (1-t)S_2 + tS_1 S_2 = [(1-t)I_n + tS_1]\, S_2, \\
d(t) &= \det M(t); \ \lambda(t) = \text{smallest eigenvalue of } M(t).
\end{aligned}
$$

1°) Show that $M(t)$ is always nonsingular for $0 \leqslant t \leqslant 1$.

2°) (a) Explain why λ is a continuous function.

(b) Use this property and the result of the question above to show that $\lambda(1) > 0$.

(c) Deduce that $S_1 S_2$ is positive definite.

---

312. ★★★ Minimizing a quadratic function over a convex polyhedron

The problem we are considering is that of minimizing a quadratic function $f : \mathbb{R}^n \to \mathbb{R}$ defined by

$$f(x) = x^T A x + b^T x + c,$$

where $A \in \mathcal{M}_n(\mathbb{R})$ is symmetric, $b \in \mathbb{R}^n, c \in \mathbb{R}$, over an unbounded closed convex polyhedron C of $\mathbb{R}^n$, defined for example as:

$$C = \left\{ x \mid a_i^T x \leqslant \beta_i \text{ for } i = 1, ..., m \right\},$$

where $a_i \in \mathbb{R}^n$ and $\beta_i \in \mathbb{R}$.

Since C is assumed unbounded, f can be bounded or unbounded from below on C.

Suppose that f is bounded from below on C; let $\inf_C f$ denote the infimum (or greatest lower bound) of f on C.

Then prove that f is minimized at some point in C, that is to say: there exists an x^* in C for which

$$f(x^*) = \inf_C f.$$

---

313. ★★★ *The convex hull of the set of orthogonal matrices*

Let $\mathcal{O}_n(\mathbb{R})$ denote the set of orthogonal matrices in $\mathcal{M}_n(\mathbb{R})$. We equip $\mathcal{M}_n(\mathbb{R})$ with two norms:

$$\nu_1(M) = \sqrt{\operatorname{tr}(M^T M)},$$
$$\nu_2(M) = \sup_{x \neq 0} \frac{\|Mx\|}{\|x\|},$$

where $\|\cdot\|$ stands for the usual Euclidean norm in $\mathbb{R}^n$.

1°) (a) By writing

$$[\nu_2(M)]^2 = \sup_{x \neq 0} \left(\frac{\|Mx\|}{\|x\|} \right)^2 = \sup_{x \neq 0} \frac{x^T(M^T M)x}{\|x\|^2},$$

check that $\nu_2(M)$ is the square root of the largest eigenvalue of the positive semidefinite matrix $M^T M$.

(b) Check that $\nu_1(M)$ is the square root of the sum of all eigenvalues of the positive semidefinite matrix $M^T M$.

From now on, we denote by $\sigma_1(M) \geqslant \sigma_2(M) \geqslant ... \geqslant \sigma_n(M) \geqslant 0$ the square roots of the eigenvalues $\lambda_1(M^T M) \geqslant \lambda_2(M^T M) \geqslant ... \geqslant \lambda_n(M^T M) \geqslant 0$ of the positive semidefinite matrix $M^T M$. Hence, it results from the above that

$$\nu_1(M) = \sqrt{\lambda_1(M^T M) + \lambda_2(M^T M) + ... + \lambda_n(M^T M)} \tag{1}$$
$$= \sqrt{\sigma_1(M)^2 + \sigma_2(M)^2 + ... + \sigma_n(M)^2},$$
$$\nu_2(M) = \sigma_1(M). \tag{2}$$

2°) Check that $\mathcal{O}_n(\mathbb{R})$ is a compact set contained in the sphere (of $\mathcal{M}_n(\mathbb{R})$) centered at the origin and of radius $\sqrt{n}$, for the ν_1 norm.

3°) Let S denote the convex hull of $\mathcal{O}_n(\mathbb{R})$. It results from the previous question that S is a compact convex set (of $\mathcal{M}_n(\mathbb{R})$) included in the (closed) ball centered at the origin and of radius $\sqrt{n}$, for the ν_1 norm.

$4°$) (a) Prove that the extreme points of S are precisely the matrices in $\mathcal{O}_n(\mathbb{R})$.

(b) Prove that S is exactly the unit ball (of $\mathcal{M}_n(\mathbb{R})$) centered at the origin, for the ν_2 norm:

$$co\left[\mathcal{O}_n(\mathbb{R})\right] = \{M \in \mathcal{M}_n(\mathbb{R}) : \sigma_1(M) \leqslant 1\}. \tag{3}$$

---

314. ★★★ *(Almost) nonnegative polynomials are (almost) sums of squares*

$1°$) If a real polynomial function P is the sum of squares of two other polynomial functions A and B, that is to say $P = A^2 + B^2$, it is clear that $P(x) \geqslant 0$ for all $x \in \mathbb{R}$. What about the converse: if P is a polynomial function which is nonnegative on $\mathbb{R}$, could it be written as the sum of squares of two polynomial functions?

$2°$) A first weakening. Suppose that the polynomial function P satisfies $P(\mathbb{R}^+) \subset \mathbb{R}^+$ (*i.e.*, $P(x)$ is nonnegative for nonnegative x). Prove that there are two polynomial functions A and B such that

$$P = A^2 + xB^2. \tag{1}$$

$3°$) A second weakening. Suppose that the polynomial function P only satisfies $P([-1, 1]) \subset \mathbb{R}^+$. Prove that there are two polynomial functions A and B such that

$$P = A^2 + (1 - x^2)B^2. \tag{2}$$

---

315. ★★★ *Extension of* MARDEN's *theorem to parallelograms*

Let the four complex numbers z_1, z_2, z_3, z_4 form the vertices of a non-degenerate parallelogram Π in the plane.

$1°$) (a) Show that there exists one, and only one, ellipse $\mathcal{E}$ of largest area contained in Π.

(b) Check that $\mathcal{E}$ is tangent to the four sides of Π exactly at the midpoints of these sides.

$2°$) Let $P(z) = (z - z_1)(z - z_2)(z - z_3)(z - z_4)$ and consider the three roots of the derivative polynomial P' of P.

Prove that one of these three roots is the center of $\mathcal{E}$ and the other two are the foci of $\mathcal{E}$.

---

316. ★★★ *A strange decomposition of real polynomial functions as sums of periodic functions*

Let P be a real polynomial function of degree $n \geqslant 1$.

$1°$) Show that P cannot be written as the sum of at most n periodic functions.

$2°$) Show that P can always be written as the sum of $n + 1$ periodic functions.

---

317. ★★★ *A* MONGE–AMPÈRE *type equation*

Let $f : (x, y) \in \mathbb{R}^2 \mapsto f(x, y) \in \mathbb{R}$ be a function that we assume to be convex and of class $\mathcal{C}^2$. We suppose that

$$\det \left[\nabla^2 f(x, y) \right] = 1 \text{ for all } (x, y) \text{ in } \mathbb{R}^2, \tag{1}$$

where $\nabla^2 f(x, y)$ denotes the Hessian matrix (or matrix of partial second derivatives) of f at (x, y).

Prove that f is necessarily a quadratic function.

---

318. ★★★ *An eikonal type equation*

Let $f : \mathbb{R}^n \to \mathbb{R}$ be a differentiable function. Assume that

$$\|\nabla f(x)\| = 1 \text{ for all } x \in \mathbb{R}^n, \tag{1}$$

where $\|\cdot\|$ denotes the usual Euclidean norm on $\mathbb{R}^n$.

What can be said about f?

---

319. ★★★ *Positive polynomials vs sums of squares*

Consider the following homogeneous polynomial of degree six in three variables x, y, z:

$$P(x, y, z) = z^6 + x^4 y^2 + x^2 y^4 - 3x^2 y^2 z^2.$$

Prove that $P(x, y, z) \geqslant 0$ for all (x, y, z) in $\mathbb{R}^3$, but that P cannot be written as the sum of squares of other (real) polynomials.

---

320. ★★★ *Images of two quadratic forms*

Consider an integer $n \geqslant 2$ and $\mathbb{R}^n$ equipped with the usual Euclidean norm $\|\cdot\|$. Let A and B be two real symmetric $n \times n$ matrices. We set

$$K = \left\{ \left(x^T A x, x^T B x \right) \mid x \in \mathbb{R}^n, \|x\| = 1 \right\}.$$

K is a compact connected subset of $\mathbb{R}^2$.

1°) Determine K in the following particular example:

$$n = 2, \quad A = \begin{bmatrix} 1 & 0 \\ 0 & -1 \end{bmatrix}, B = \begin{bmatrix} 0 & 1 \\ 1 & 0 \end{bmatrix}.$$

2°) Suppose that $n \geqslant 3$. Prove, simply if possible, that K is convex.

321. ★★★ *Arcwise-connectedness vs convexity*

A closed convex set C in $\mathbb{R}^n$ is necessarily arcwise-connected and satisfies the following property:

$$\left(x^T d \in \{c^T d \mid c \in C\} \text{ for all } d \in \mathbb{R}^n\right) \Rightarrow (x \in C). \tag{1}$$

Question: Is a closed arcwise-connected set C satisfying (1) convex?

---

322. ★★★ *Partitions of $\mathbb{R}^n$ with convex sets*

Let $C_1, C_2, \ldots, C_p$ be convex sets of $\mathbb{R}^n$ forming a partition of $\mathbb{R}^n$.

Are the closed convex sets $\overline{C_1}, \overline{C_2}, \ldots, \overline{C_p}$ necessarily polyhedral?

---

323. ★★★ *Sets of matrices vs their images*

1°) The countable case.

Let $\mathcal{A}$ be a countable set in $\mathcal{M}_{m,n}(\mathbb{R})$; we set $\mathcal{A}x = \{Ax \mid x \in \mathbb{R}^n\}$, that is the set (in $\mathbb{R}^m$) of images of $x \in \mathbb{R}^n$ by the matrices A in $\mathcal{A}$.

Does $B \in \mathcal{M}_{m,n}(\mathbb{R})$ satisfying

$$Bx \in \mathcal{A}x \text{ for all } x \in \mathbb{R}^n \tag{1}$$

necessarily belong to $\mathcal{A}$?

2°) The convex compact case.

Suppose here that $\mathcal{A} \subset \mathcal{M}_{m,n}(\mathbb{R})$, instead of being countable, is convex and compact. We then pose the same question as above.

---

324. ★★★ *Continuity of a mapping by compactness and connectedness of images*

1°) Find a (simple) discontinuous mapping $f : \mathbb{R}^n \to \mathbb{R}^m$ such that the image by f of any compact subset of $\mathbb{R}^n$ is a compact subset of $\mathbb{R}^m$.

2°) Find a discontinuous function $f : \mathbb{R} \to \mathbb{R}$ such that the image by f of any interval is an interval.

3°) Consider a mapping $f : \mathbb{R}^n \to \mathbb{R}^m$ such that the image by f of any compact subset is a compact subset and the image by f of any connected subset is a connected subset.

Prove that f is necessarily continuous.

---

325. ★★★ *A nice convex function of matrices and vectors*

Let $\mathcal{P}_n$ denote the set of $n \times n$ symmetric positive definite matrices. It is an open convex cone of the vector space $\mathcal{S}_n(\mathbb{R})$ of (n, n) symmetric real matrices. Let now $f : \mathcal{P}_n \times \mathbb{R}^n \to \mathbb{R}$ be defined as:

$$f(A, x) = x^T A^{-1} x. \tag{1}$$

The convexity of the partial function $f(A, \cdot)$ is well-known (it is just a quadratic form on $\mathbb{R}^n$ associated with a positive definite matrix A^{-1}). The convexity of the partial function $f(\cdot, x)$ is also secured; indeed, the inverting operation $(\cdot)^{-1} : \mathcal{P}_n \to \mathcal{P}_n$ which assigns A^{-1} to A is convex; in saying that, we mean that $\mathcal{P}_n$ is equipped with the order

$$M \preccurlyeq N \text{ if and only if } x^T M x \leqslant x^T N x \text{ for all } x \in \mathbb{R}^n$$

and that

$$(\alpha M + \beta N)^{-1} \preccurlyeq \alpha M^{-1} + \beta N^{-1} \text{ whenever } \alpha > 0, \beta > 0, \alpha + \beta = 1.$$

The question we pose now is: is $f(\cdot, \cdot)$ a (jointly) convex function on $\mathcal{P}_n \times \mathbb{R}^n$?

---

326. ★★★ *Slices in a spherical bread*

Before going to university, a student takes breakfast with slices in a spherical loaf of bread. This spherical loaf, of diameter $2r$, is sliced into n pieces of equal thickness $h = 2r/n$.

1°) Which slice, with cuts at a and $a + h$, has the most bread?

2°) Which slice, with cuts at a and $a + h$, has the most crust?

---

327. ★★★ *Behavior of the volume of unit balls and of the surface area of unit spheres*

Consider $\mathbb{R}^n$ equipped with the usual Euclidean norm $\|\cdot\|$; let V_n denote the volume of the unit ball, let S_n denote the surface area of the unit sphere.

1°) (a) What are the values of V_n for n up to 6?

(b) What is the general expression of V_n?

(c) Using a relationship between V_{n+2} and V_n, determine the value(s) of n for which V_n is maximal.

2°) Show that the series with general term V_n is convergent and calculate its sum

$$\sum_{n=1}^{+\infty} V_n.$$

3°) (a) What is the general relation linking S_n and V_n?

(b) Determine the value(s) of n maximizing S_n.

---

328. ★★★ *Two basic inequalities concerning a triangle*

1°) Let r and R be the radiuses of the inscribed and circumscribed circles of a triangle. Then prove:

$$R \geqslant 2r, \tag{1}$$

with equality if and only if the triangle is equilateral.

97

2°) Let R_1, R_2, R_3 be the distances to the three vertices of a triangle from any interior point P of that triangle. Let r_1, r_2, r_3 be the distances from P to the three sides. Then prove:

$$R_1 + R_2 + R_3 \geqslant 2(r_1 + r_2 + r_3),\qquad (2)$$

with equality if and only if the triangle is equilateral and P is its centroid.

---

329. ★★★ *A property of normal vectors to faces of a tetrahedron*

Let $ABCD$ be a tetrahedron. Orthogonal to each of its four faces, set an outward normal vector whose length is equal to the area of the corresponding face.

Prove that the sum of these four normal vectors is the zero vector.

---

330. ★★★ *Continuity or non-continuity of the derivation operation*

Consider the real vector space $\mathcal{C}^\infty(\mathbb{R})$ and the derivation operation $(\cdot)'$ acting on it:

$$(\cdot)' : f \in \mathcal{C}^\infty(\mathbb{R}) \mapsto f' \in \mathcal{C}^\infty(\mathbb{R}).$$

1°) Let us equip $\mathcal{C}^\infty(\mathbb{R})$ with a norm $\|\cdot\|$ so that $(\mathcal{C}^\infty(\mathbb{R}), \|\cdot\|)$ is a normed vector space. Show that the derivation operation cannot be continuous on $\mathcal{C}^\infty(\mathbb{R})$.

2°) Consider now the real vector space $\mathcal{P}$ of polynomial functions on $\mathbb{R}$ (thus a subspace of $\mathcal{C}^\infty(\mathbb{R})$). We equip $\mathcal{P}$ with a norm $\|\cdot\|$ so that $(\mathcal{P}, \|\cdot\|)$ is a normed vector space. Is the derivation operation

$$(\cdot)' : f \in \mathcal{P} \mapsto f' \in \mathcal{P}$$

continuous on $\mathcal{P}$?

---

331. ★★★ *The magic of* FOURIER *series (6)* (More "calculatory")

A physical signal is represented by the following function: $f : \mathbb{R} \to \mathbb{R}$, periodic of period 2π, with:

$$f(x) = e^x \text{ for } 0 \leqslant x < 2\pi.$$

1°) Sketch the graph of f for x between -2π and 2π.

2°) Calculate the (real) FOURIER coefficients associated with f (denoted as a_n and b_n).

3°) Give a clearly argued reason why f can be written as

$$\frac{f(x+) + f(x-)}{2} = a_0 + \sum_{n=1}^{+\infty} [a_n \cos(nx) + b_n \sin(nx)] \text{ for all } x \in \mathbb{R}. \qquad (1)$$

[$f(x+)$ (resp. $f(x-)$) denotes the right-limit (resp. the left-limit) of f at x.]

$4°$) (a) Give a simple argument for the convergence of the series with general terms $u_n = \frac{1}{n^2+1}$ and $u_n = \frac{\cos n}{n^2+1}$.

(b) By choosing two particular values of x in the development (1) above, determine $\sum_{n=1}^{+\infty} \frac{\cos n}{n^2+1}$.

$5°$) Use the results above to determine $\sum_{n=1}^{+\infty} \frac{1}{n^2+1}$.

---

332. ★★★ *"Variational "characterizations of the sine function*

Prove that the sine function is the only one minimizing

$$\mathcal{I}_1(f) = \int_0^\pi \left([f'(x)]^2 - [f(x)]^2 \right) \, dx \tag{1}$$

among the set of functions $f \in \mathcal{C}^1([0, \pi], \mathbb{R})$ satisfying

$$f(0) = 0, \ f'(0) = 1, \ f(\pi) = 0. \tag{2}$$

---

333. ★★★ *Maximizing a function of eigenvalues*

Let $M \in \mathcal{S}_n(\mathbb{R})$ with eigenvalues $\lambda_1 \leqslant \lambda_2 \leqslant \ldots \leqslant \lambda_n$; let

$$\mathcal{E}(M) = \left\{ \Omega M \Omega^{-1} : \Omega \text{ is an orthogonal } (n, n) \text{ matrix} \right\}.$$

Consider a convex function $f : [\lambda_1, \lambda_n] \to \mathbb{R}$.

$1°$) Prove that if $A = [a_{ij}] \in \mathcal{E}(M)$, then

$$\lambda_1 \leqslant a_{ii} \leqslant \lambda_n \text{ for all } i = 1, 2, \ldots, n. \tag{1}$$

$2°$) Deduce from the above the value of

$$\max \left\{ \sum_{i=1}^n f(a_{ii}) : A = [a_{ij}] \in \mathcal{E}(M) \right\}.$$

Chapter 2. Hint(s)

1. *Hints.* 1°) A possible proof of (1) is by induction.

    2°) - Apply (1) with n circular permutations σ:

$$x_1y_1 + x_2y_2 + \ldots + x_ny_n \leqslant x_1y_1 + x_2y_2 + \ldots + x_ny_n;$$
$$x_1y_2 + x_2y_3 + \ldots + x_ny_1 \leqslant x_1y_1 + x_2y_2 + \ldots + x_ny_n;$$
$$\ldots$$
$$x_1y_n + x_2y_1 + \ldots + x_ny_{n-1} \leqslant x_1y_1 + x_2y_2 + \ldots + x_ny_n;$$

and sum the inequalities to obtain:

$$(x_1 + x_2 + \ldots + x_n)(y_1 + y_2 + \ldots + y_n) \leqslant n\,(x_1y_1 + x_2y_2 + \ldots + x_ny_n).$$

    - One could also prove (2) without passing through (1), with the help of the following (general) expansion:

$$\sum_{i,j=1}^{n}(x_i - x_j)(y_i - y_j) = 2n\sum_{k=1}^{n}x_ky_k - 2\left(\sum_{k=1}^{n}x_k\right)\left(\sum_{k=1}^{n}y_k\right). \tag{3}$$

    With the assumption $x_1 \leqslant x_2 \leqslant \ldots \leqslant x_n$ and $y_1 \leqslant y_2 \leqslant \ldots \leqslant y_n$, the quantity above is nonnegative.

2. *Hint.* A possibility is to prove this inequality by induction on n.

3. *Hint.* Apply, for example, the CAUCHY–SCHWARZ inequality to $a_1 = \sqrt{x_1}, \ldots, a_n = \sqrt{x_n}$ and $b_1 = 1/\sqrt{x_1}, \ldots, b_n = 1/\sqrt{x_n}$.

5. *Hint.* Make use of the CAUCHY–SCHWARZ inequality.

7. *Hint.* Isolate the terms $(n - k)(k + 1)$ in the expression for $(n!)^2$, then find lower bounds for $(n - k)(k + 1)$.

8. *Hint.* Use the expansion of $(1 - 1)^n$ in parallel with that of $(1 + 1)^n$.

9. *Hint.* 2°) Observe that $Q_n = S_n/(S_{2n} - S_n)$.

10. *Hint.* Show that $\sum_{k=1}^{n}k = \frac{n(n+1)}{2}$ and $\sum_{k=1}^{n}k^3 = \left[\frac{n(n+1)}{2}\right]^2$.

11. *Hint.* To establish (1), use the binomial expansion of $(p + 1)^{k+1}$ and sum the results from $p = 1$ to $p = n$.

100

© Springer International Publishing Switzerland 2016
J.-B. Hiriart-Urruty, *Mathematical Tapas*, Springer Undergraduate
Mathematics Series, DOI 10.1007/978-3-319-42186-5_2

14. *Hints.* - The distance between home and university will not be of importance here.
   - Beware, the answer is not 30 m/h.

17. *Hints.* 1°) If n is prime, the only divisors of n are 1 and n, whence $\sigma(n) = n + 1$.
   2°) Decompose n as $n = pq$, where $1 < q \leqslant p < n$. Thus, $\sigma(n) \geqslant 1 + p + n$.

21. *Hint.* 1°) Begin by proving, by induction or by decomposing fractions into simple elements, that

$$\sum_{n=1}^{N} \frac{1}{n(n+1)(n+2)...(n+p)} = \frac{1}{p}\left(\frac{1}{p!} - \frac{1}{N(N+1)(N+2)...(N+p)}\right).$$

22. *Hint.* 2°) Observe that to have

$$(u, v) \in \mathbb{Q}^2 \ ; \ u > 0, v > 0 \text{ and } u^n + v^n = 1 \tag{1}$$

is equivalent to having

$$(x, y, z) \in \mathbb{N}^3 \ ; \ x > 0, y > 0, z > 0 \text{ and } x^n + y^n = z^n. \tag{2}$$

This is precisely FERMAT's equation.

23. *Hint.* It suffices to consider the case where $|x - y| \geqslant \frac{1}{2}$. Then use the triangle inequality on $|\cdot|$.

26. *Hint.* A possibility is to use the mean value theorem for the function

$$x \in [0, 1] \mapsto F(x) = \int_0^x f(t) \, dt - \frac{x}{2}.$$

30. *Hint.* A function satisfying (1) is necessarily even. For any $a > 0$, consider the sequence $(x_n)_n$ defined as follows:

$$\begin{aligned} x_0 &= a; \\ \text{for all } n &\geqslant 0, \ x_{n+1} = \sqrt{x_n}. \end{aligned}$$

We then have: $\lim_{n \to +\infty} x_n = 1$ and $f(a) = f(x_1)... = f(\lim_{n \to +\infty} x_n)$. Whence $f(a) = f(1)$.

31. *Hint.* Use the change of variables $u = \pi - x$ in the integral $\int_0^{\frac{\pi}{2}} x f(\sin x) \, dx$.

33. *Hint.* 2°) Try with $Q_\varepsilon(x) = P_n(x^2)$.

34. *Hint.* Consider the primitive (or antiderivative) function $F(x) = \int_0^x f(t) \, dt$, then the function $g(x) = e^{-x} F(x)$. Study the variations of the function g.

101

37. *Hints.* 3°) Either use the definition of the arctan function, or differential calculus ($f' = 0$ on an interval implies that f is constant on it).

38. *Hints.* 1°) Use differential calculus: the functions in the two sides of (1) are continuous on $[0, 1]$, differentiable on $(0, 1)$, and their derivatives are equal.

    2°) Check that $f'(x) = 0$ for all $x \in \mathbb{R}$.

39. *Hint.* Pass via the function $\log(f)$.

46. *Hint.* Use the second-order TAYLOR–YOUNG series of f around a.

49. *Hint.* 2°) $S_n \geqslant 0$ for all n and the sequence (S_n) is decreasing; it is therefore convergent. Consequently

$$S_n - S_{n-1} = -\frac{2}{\pi} b_n^2 \to 0 \text{ as } n \to +\infty.$$

50. *Hint.* Apply the intermediate value theorem to the continuous function $g : [0, 1 - \frac{1}{n}] \to \mathbb{R}$ defined by $g(x) = f\left(x + \frac{1}{n}\right) - f(x)$.

52. *Hint.* Use the intermediate value property or ROLLE's theorem.

53. *Hint.* 1°) Use the mean value theorem.

54. *Hint.* Apply the mean value theorem to the function $\log f$.

55. *Hint.* Apply the mean value theorem to the function $g(x) = e^{-kx} f(x)$.

58. *Hint.* Begin by reordering the $a_i's : a_{i_1} < a_{i_2} < ... < a_{i_n}$. Then distinguish two cases: when n is odd and when n is even.

61. *Hint.* 1°) and 2°): show that $h'_1 = 0$ and $h'_2 = 0$ on $\mathbb{R}$.

65. *Hint.* As often in such a situation, use complex numbers: the affixes of the n vertices of the polygon are the n-th roots of 1. Then, recognize in the mean value of the lengths of the chords a RIEMANN sum associated with the integral $\int_0^1 \sin(\pi x)\, dx$.

66. *Hint.* Identifying the plane with $\mathbb{C}$, the n points S_k, $k = 1, 2, ..., n$, have $r_k = \exp\left(i \frac{2\pi(k-1)}{n}\right)$ as affixes. What is asked for is the value of $\prod\limits_{k=2}^{n} |1 - r_k|$.

    It is therefore useful to consider the polynomial function $P_{n-1}(z) = 1 + z + z^2 + ... + z^{n-1} \left(= \frac{z^n - 1}{z - 1} \text{ if } z \neq 1\right)$ whose roots are $r_2, r_3, ... , r_n$.

67. *Hint.* Check that x, y, z are necessarily the roots of a polynomial of degree 3.

102

70. *Hint.* Express the determinant of $M = [m_{i,j}]$ as a sum of products of random variables whose expectations are all null (because the expectation of $m_{i,j}$ is 0 and the expectation of the product of independent random variables is the product of their expectations).

72. *Hints.* 1°) Just use the two basic calculus rules: $\|u + v\|^2 = \|u\|^2 + \|v\|^2 + 2u^T v$ and $\|\lambda u\|^2 = \lambda^2 \|u\|^2$.

   2°) (*b*) If $A(x, y) = 0$, then

$$\|y\|^2 x - (x^T y)\, y = 0 \ \text{(from (1))};$$
$$|x^T y| = \|x\| \times \|y\| \ \text{(from (2))},$$

whence one obtains $\frac{x}{\|x\|} = \pm \frac{y}{\|y\|}$, which means that x and y are collinear vectors. This is "the equality case" in the CAUCHY–SCHWARZ inequality.

74. *Hint.* Have the "optimization reflex" by considering $h(x) = f(x) - g(x)$. For such a function, a is a local minimizer or a local maximizer.

75. *Hints.* $[(ii) \Rightarrow (iii)]$ is clear.

   Indications for proving $[(iii) \Rightarrow (i)]$. One shows firstly that $\left| \frac{b^2 + c^2 - a^2}{2bc} \right| < 1$; hence there exists an angle $\widehat{A} \in (0, \pi)$ such that $\cos \widehat{A} = \frac{b^2 + c^2 - a^2}{2bc}$. One therefore constructs a triangle based on this angle $\widehat{A}$ and with two sides AB and AC of lengths b and c, respectively. The length of the third side BC of such a triangle is necessarily a' satisfying $\cos \widehat{A} = \frac{b^2 + c^2 - a'^2}{2bc}$; consequently $a = a'$.

77. *Hint.* 1°) Consider the graph of the function $x > 0 \mapsto 1/x$ and use the fact that $\int_1^n \frac{1}{t}\,dt = \ln(n)$.

78. *Hint.* Write the inequality (1) for $x = n, n + 1, ..., pn - 1$ and add up the resulting inequalities.

79. *Hint.* Use the following result:

$$\sum_{i=1}^k \frac{1}{i} - \ln k \text{ has a limit } \gamma \text{ as } k \to +\infty.$$

80. *Hint.* 2°) Consider n as "squeezed" between two consecutive powers of 2, *i.e.*, $2^p \leqslant n < 2^{p+1}$. Then prove, by induction on n, that $H_n = \frac{1}{2^p} \frac{q}{r}$ with odd integers q and r.

82. *Hint.* Without loss of generality, one can assume that $x \geqslant y \geqslant z$.

86. *Hint.* The two questions 1°) and 2°) are independent.

103

89. *Hint.* Use the inequality

$$n! \leqslant \sum_{k=1}^{n} k! \leqslant 2(n-1)! + n!$$

90. *Hint.* Use the auxiliary sequence defined by $v_n = \frac{u_n}{(2n)!}$.

91. *Hint.* Use the following decomposition:

$$\text{For } n \geqslant 2, \quad \frac{u_n}{n} = \frac{u_1}{n} + \frac{v_1 + v_2 + \dots + v_{n-1}}{n}.$$

94. *Hint.* Only the implication $[\impliedby]$ asks for some effort. For that, given $\varepsilon > 0$, add the following inequalities, valid for n large enough:

$$l - \varepsilon \;\leqslant\; 2u_{n+1} - u_n \leqslant l + \varepsilon;$$
$$2(l-\varepsilon) \;\leqslant\; 4u_{n+2} - 2u_{n+1} \leqslant 2(l+\varepsilon);$$
$$\dots$$
$$2^{k-1}(l-\varepsilon) \;\leqslant\; 2^k u_{n+k} - 2^{k-1}u_{n+k+1} \leqslant 2^{k-1}(l+\varepsilon).$$

A prototype of sequence (u_n) satisfying the property in the right hand-side of (1) is: $u_{n+1} = \frac{u_n + l}{2}$. This observation may be used to provide another proof of the implication $[\impliedby]$.

95. *Hint.* In the calculation of the sum of the series, use the following result (see Tapa 78): given an integer $p \geqslant 2$,

$$\text{the limit of } S_p(N) = \sum_{n=N}^{pN} \frac{1}{n} \text{ as } N \to +\infty \text{ is } \log(p).$$

96. *Hint.* 2°) The left-hand side in (1) is a RIEMANN sum for the integral $\int_0^1 \frac{dx}{1+x}$.

99. *Hint.* 2°) The following relation will be useful

$$\left[\prod_{k=1}^{n} \cos\left(\frac{\theta}{2^k}\right)\right] \times \sin\left(\frac{\theta}{2^n}\right) = \frac{\sin(\theta)}{2^n}. \tag{1}$$

This can be checked with the help of the rule $\sin(2\alpha) = 2\sin(\alpha)\cos(\alpha)$.

109. *Hint.* A proof by induction is suggested.

110. *Hint.* The solution of $(\mathcal{C}_3)$ is defined on the whole $\mathbb{R}$. To find it, "glue together" pieces of solutions of the first two differential equations $(\mathcal{C}_1)$ and $(\mathcal{C}_2)$.

104

113. *Hint.* 1°) Pass to polar coordinates.

114. *Hint.* Start with the inequality

$$2\,|f'(x)|\,\sqrt{|f(x)|} \leqslant |f(x)| + |f'(x)|^2\,,$$

and apply a change of variables to the (convergent) integral $\int_0^{+\infty} |f'(x)|\,\sqrt{|f(x)|}\,dx$.

115. *Hint.* Decompose the polynomial functions in a chosen basis consisting of LAGRANGE polynomial functions.

116. *Hint.* Begin with polynomial functions. Then use the fact that f can be approximated uniformly on $[0,1]$ by polynomial functions (WEIERSTRASS approximation theorem).

117. *Hint.* Check firstly that $\int_0^1 \left(x - \frac{1}{2}\right)^n f(x)\,dx = 1$. Then prove that the assumption "$|f(x)| < (n+1)2^n$ for all $x \in [0,1]$" leads to a contradiction.

118. *Hint.* Use the following facts:
   - If r_i is a root (of P) of multiplicity m_i, then $r_1^{m_1} \times r_2^{m_2} \times \ldots \times r_k^{m_k} = \pm 1$.
   - $P(0) = -1$, and $P(x) \to +\infty$ as $x \to +\infty$.

123. *Hint.* 3°) The image of a connected subset of $\mathbb{R}^2$ (hence of the convex subset A) by a real-valued continuous function like g is an interval of $\mathbb{R}$.

124. *Hint.* 1°) Start with a function F_1 defined as:

$$F_1(x) = \begin{cases} x^2 \sin\left(\frac{1}{x}\right) & \text{if } x \neq 0, \\ 0 & \text{if } x = 0. \end{cases}$$

2°) If both f^2 and g^2 were derivative functions, that would also be the case for their sum; but $f^2 + g^2$ cannot be a derivative function since it does not satisfy the intermediate value property.

128. *Hints.* As often happens in mathematics, the inequality (1) is easy to derive if one deals with it in the right way. Firstly, the graph of f on $[a,b]$ is under the chord joining the end-points $(a, f(a))$ and $(b, f(b))$, that is:

$$\text{For all } x \in [a,b],\ f(x) \leqslant f(a) + \frac{f(b) - f(a)}{b - a}(x - a). \tag{2}$$

Secondly, decompose $\int_a^b f(x)\,dx$ as $\int_a^{\frac{a+b}{2}} f(x)\,dx + \int_{\frac{a+b}{2}}^b f(x)\,dx$.

130. *Hint.* Begin by proving that f is of class $\mathcal{C}^\infty$. Then take derivatives with respect to x and with respect to y in the functional relation (1).

131. *Hint.* Consider $g = f'$. This new function then solves a first-order differential equation.

143. *Hint.* A proof by contradiction is suggested.

144. *Hint.* 1°) Consider the derivative of the function $x \mapsto \int_x^{x+T} f(t)\, dt$.

146. *Hint.* One way (but it is not the only one) is to use the CAUCHY–SCHWARZ inequality.

147. *Hint.* Use first-order and second-order TAYLOR series of f around a point.

150. *Hint.* Use the following identity: $(x * x) * (y * y) = (x * y) * (x * y)$.

151. *Hints.* 1°) Consider p such that $(e - x)^p = 0$ and expand $(e - x)^p$ by using the binomial formula.

2°) Nilpotence of $x + y$ and $x - y$. Consider p such that $x^p = y^p = 0$. Then expand $(x + y)^{2p-1}$ by using the binomial formula.

153. *Hint.* Use the expressions of the areas of the subtriangles PAB, PAC, PBC and then the arithmetic-geometric inequality.

157. *Hint.* Observe that if d is the number of elements of G, then $A^d = I_n$ for any element A in G (this is a consequence of LAGRANGE's theorem in group theory). Consequently the minimal polynomial of A is a divisor of the polynomial $X^d - 1$; hence all its roots are simple, and A is diagonalizable.

158. *Hint.* Firstly observe that $\mathcal{V}$ is contained in the kernel of the linear mapping

$$\text{trace} : \mathcal{M}_n(\mathbb{R}) \to \mathbb{R}.$$

Secondly, find as many linearly independent matrices as possible in $\mathcal{V}$, built up from the basic matrices $E_{i,j}$ (with 1 as entry (i,j), 0 elsewhere).

160. *Hint.* Call on matrices of the form $M - \alpha I_n$ where $\alpha \in \mathbb{C}$ is not an eigenvalue of M.

161. *Hints.* Resources either from Linear algebra or from Analysis can be used.

- Using Linear algebra. A proof by contradiction is suggested: suppose that all matrices in $\mathcal{H}$ are singular, develop an argument arriving at a contradiction. $\mathcal{H}$ is the kernel of a non-null linear form $u : \mathcal{M}_n(\mathbb{R}) \to \mathbb{R}$. Since I_n is assumed not to belong to $\mathcal{H}$, we may assume, without loss of generality, that $u(I_n) = 1$ (if not, divide u by the non-null $u(I_n)$). The objective is to bring out two matrices in $\mathcal{H}$ (they are singular matrices) whose sum in not in $\mathcal{H}$ (it is an invertible matrix).

- Using the Euclidean structure of $\mathcal{M}_n(\mathbb{R})$. As usual, $\mathcal{M}_n(\mathbb{R})$ is equipped with the inner product $(U, V) \mapsto \langle\langle U, V \rangle\rangle = \text{tr}(U^T V)$. Hence, the hyperplane $\mathcal{H}$ can be represented as follows: there exists a non-null $H \in \mathcal{M}_n(\mathbb{R})$ such that

$$\mathcal{H} = \{ M \in \mathcal{M}_n(\mathbb{R}) : \langle\langle H, M \rangle\rangle = 0 \}.$$

106

Given the non-null $H = [h_{ij}]$, the aim is to design an invertible $M = H = [m_{ij}]$ such that $\langle\langle H, M \rangle\rangle = \text{tr}(H^T M) = \sum_{i,j} h_{ij} m_{ij} = 0$.

- Using Analysis. A proof by contradiction is suggested: suppose that all matrices in $\mathcal{H}$ are singular, and develop an argument arriving at an impossibility. $\mathcal{H}$ is the kernel of a non-null linear form $u : \mathcal{M}_n(\mathbb{R}) \to \mathbb{R}$. Let $\Omega \subset \mathcal{M}_n(\mathbb{R})$ denote the set of invertible matrices. Then: Ω is an open dense subset in $\mathcal{M}_n(\mathbb{R})$ (see Tapa 160); it is made of two connected components, $\Omega_1 = \{M \in \Omega : \det M > 0\}$ and $\Omega_2 = \{M \in \Omega : \det M < 0\}$. In view of the assumption made, we necessarily have (within a permutation ± 1):

$$\Omega_1 \subset \mathcal{H}_+ = \{M : u(M) > 0\} \text{ and } \Omega_2 \subset \mathcal{H}_- = \{M : u(M) < 0\}.$$

There are singular matrices in $\mathcal{H}_+$ and in $\mathcal{H}_-$ (because the set of singular matrices cannot equal the vector space $\mathcal{H} = \{M : u(M) = 0\}$). Consider, for example, a singular matrix A in $\mathcal{H}_+$ and design a sequence of matrices (B_n) in $\Omega_2 \subset \mathcal{H}_-$ converging towards A.

165. *Hint.* The expression of Δ_n in a closed form involves a power of f (denoted as f^p) and a higher order derivative of $1/f$ (denoted as $(1/f)^{(q)}$).

168. *Hints.* 2°) - Check that if M_σ and M_τ are two permutations matrices, then $M_\sigma M_\tau = M_{\sigma \circ \tau}$.

- Consequence: since any permutation is the composition of a finite number of transpositions, it suffices that A commutes with the (specific) permutation matrices associated with transpositions.

169. *Hints.* Use calculus rules governing products of matrices which are triangular by blocks.

1°) (a) Multiply M by $\begin{bmatrix} D^T & 0 \\ C^T & D^{-1} \end{bmatrix}$, and use calculus rules on determinants.

171. *Hint.* Apply the CAYLEY–HAMILTON theorem in $\mathcal{M}_2(\mathbb{C})$, so that in our specific case:

$$A^{-1} = -A + \text{tr}(A)\, I_2; \quad B^{-1} = -B + \text{tr}(B)\, I_2; \quad (AB)^2 = \text{tr}(AB)\, AB - I_2.$$

Then, use the decomposition of BA as

$$BA = A^{-1}(AB)^2 B^{-1}.$$

172. *Hints.* 1°) Show that the unique solution x of $(A^{-1} + I_n)x = 0$ is $x = 0$.

2°) Begin, because it is easier, with the case where A is diagonalizable.

173. *Hint.* In the third question, the uniqueness result is a bit more difficult to prove than the existence.

174. *Hints.* There are (at least) two ways of proving (1):

- First proof: show that $A^{2n+1} = (-r^2)^n A$ for all integers n; then $A^{2n+2} = A(A^{2n+1})$ for all n; finally, use the definition of $\exp(A)$ as the sum of a series.

- Second proof: diagonalize A in $\mathcal{M}_3(\mathbb{C})$, its eigenvalues are $0, ir, -ir$; then use the fact that $\exp(P^{-1}AP) = P^{-1}\exp(A)P$.

176. *Hint.* The inertia of q_1 (or of the associated symmetric matrix Q_1) is a triple of integers: the number of positive eigenvalues, the number of negative eigenvalues, and the number of null eigenvalues.

179. *Hint.* Let $M_\lambda = \lambda A + (1 - \lambda)B$. Consider the following sets

$$
\begin{aligned}
I_A &= \left\{ \lambda \in [0,1] \; : \; x^T M_\lambda x \geqslant 0 \text{ for all } x \in S \right\}, \\
I_B &= \left\{ \lambda \in [0,1] \; : \; x^T M_\lambda x \geqslant 0 \text{ for all } x \in T \right\}.
\end{aligned}
$$

Check that I_A and I_B are nonempty closed intervals. Then prove that $I_A \cup I_B = [0,1]$; this will imply that $I_A \cap I_B$ is nonempty, which is the aimed conclusion.

180. *Hint.* [$(a) \implies (b)$]. Prove it by contradiction. Assuming the contrary of (b), for every positive integer n, there exists an $x_n \neq 0$ such that

$$
x_n^T Q x_n + n \left\| A x_n \right\|^2 \leqslant 0,
$$

or, by putting $u_n = x_n / \left\| x_n \right\|$,

$$
\frac{u_n^T Q u_n}{n} + \left\| A u_n \right\|^2 \leqslant 0.
$$

One can then extract from (u_n) a convergent subsequence $(u_{n_k})_k$.

182. *Hints.* 1°) $\widehat{q}$ is the "homogenized" version of q, that is:

$$
\widehat{q}(x,t) = t^2 q\left(\frac{x}{t} \right).
$$

2°) $\widehat{A} \in \mathcal{S}_{n+1}(\mathbb{R})$, the "augmented" version of A, is the matrix representing the quadratic form $\widehat{q}$ on $\mathbb{R}^{n+1}$.

184. *Hints.* Guiding thread for a proof. For t in $\mathbb{R}$, let $A(t) = A + tB$. Then:

- For $t \geqslant 0$, the symmetrized product of $A(t)$ and B is positive definite.

- There exists a threshold $t_0 \geqslant 0$ such that $A(t)$ is positive definite for all $t > t_0$.

- If one supposes that $A(0) = A$ is not positive definite, show that there exists a $t_1 \in [0, t_0]$ such that $A(t_1)$ is positive semidefinite but not positive definite.

- Obtain a contradiction by using a non-null vector in the kernel of $A(t_1)$.

189. *Hints.* - For any nonsingular matrix P, the matrix PP^T is symmetric positive definite.

- A symmetric positive definite S can be factorized as $S = P^T P$, where P is a nonsingular matrix. A real symmetric matrix B is always diagonalizable.

190. *Hints.* For $x \neq 0$ in $\mathbb{R}^n$, consider the vector $y = (A^T)^{-1} x$ and the matrix $M = xy^T$. This particular matrix M is of rank 0 or 1. This allows us to determine the eigenvalues of M. Then observe that:

$$
\begin{aligned}
\det(M + I_n) &= 1 + x^T y; \\
M + I_n &= (xx^T + A)A^{-1}; \\
x^T y &= x^T A^{-1} x.
\end{aligned}
$$

191. *Hints.* Consider $(A^{-1})_x$ built up like A_x but with A^{-1} instead of A. Then calculate the product $A_x.(A^{-1})_x$ and use the rule $\det[A_x.(A^{-1})_x] = \det(A_x)\det[(A^{-1})_x]$.

192. *Hints.* 1°) Proof by contradiction, for example. Suppose that the rank of A is $r \geqslant 1$. There exist nonsingular matrices P and Q such that

$$
A = P \begin{bmatrix} I_r & 0 \\ 0 & 0 \end{bmatrix} Q.
$$

Then choose $A = P \begin{bmatrix} 0 & 0 \\ 0 & I_{n-r} \end{bmatrix} Q$ in order to contradict the relation (1).

2°) The right-hand side of (2) equivalently says that

$$
\det(A - B + Y) = \det(Y) \text{ for all } Y \text{ in } \mathcal{M}_n(\mathbb{R}).
$$

193. *Hints.* - Various proofs are possible: one can use resources either of Linear algebra or of Analysis.

- One (indirect) way would consist in using formula (1) in Tapa 202, followed by the characterization of $AB = 0$ in Tapa 309.

196. *Hints.* - Make use of the characterization of positive semidefinite matrices A in terms of minors of A.

- More specifically, for answering the first question, show that $a_{i,j} \in [-1, 1]$ for all i, j.

197. *Hint.* The results have to do with the largest and the smallest eigenvalues of A.

198. *Hint.* The results have to do with the largest and the smallest eigenvalues of A.

203. *Hint.* Proof by induction on n.

206. *Hint.* Prove that the second differential of F is null everywhere, whence F is an affine function.

207. *Hint.* Equip $\mathcal{M}_n(\mathbb{R})$ with the inner product $\langle U, V \rangle = \mathrm{tr}(U^T V)$ and the resulting norm $\|M\| = \sqrt{\langle M, M \rangle}$. Expand $\|A - B\|^2$.

208. *Hint.* $2°)$ $\|\cdot\|_{(k)}$ compares easily with $\|\cdot\|_\infty$ and $\|\cdot\|_1$; beware, however, that the string of inequalities between these norms and the string of inclusions on the corresponding unit balls are not in the same order.

209. *Hint.* Use the relation

$$\alpha\beta = \frac{1}{4}\left[(\alpha+\beta)^2 - (\alpha-\beta)^2\right]$$

with $\alpha = u_i$ and $\beta = v_i$ in order to bound $(u_i v_i)^2$. Furthermore, the assumption $u^T v \geqslant 0$ means that $\|u - v\| \leqslant \|u + v\|$.

210. *Hint.* A proof by induction on n.

212. *Hint.* To prove it, use the so-called divergence theorem (or GREEN–OSTROGRADSKY, or OSTROGRADSKY–GAUSS theorem) relating the flow of a vector field $\overrightarrow{F}$ through the boundary of S (thus, a surface integral) to a triple integral of its divergence function $div\,\overrightarrow{F}$ on S. Here, the vector-valued function to be considered is simply $\overrightarrow{F}(x, y, z) = \frac{1}{3}(x, y, z)$.

214. *Hints.* Note that $Q - Q' = P$. Prove that $\lim_{|x|\to+\infty} Q(x) = +\infty$ and $\inf_{x\in\mathbb{R}} Q(x) \geqslant 0$.

215. *Hint.* For question $3°)$, distinguish the roots of P' which are roots of P (nothing to prove in that case) and those which are only roots of P'. The key then is to decompose the fractional expression P'/P into simple elements:

$$\frac{P'(x)}{P(x)} = \sum_{i=1}^{k} \frac{b_i}{x - a_i},$$

and write it for $x = a$ root of P' which is not a root of P.

216. *Hint.* With P standing for the point with coordinates (x, y), consider the function g defined on S by $g(P) = f(P) - f(-P)$.

217. *Hint.* [$\Longleftarrow$] Prove this implication by contradiction. The key property of f to be used is its continuity.

220. *Hints.* $2°)$ Use the basic definitions of minimizers of functions and the triangle inequality on $\|\cdot\|$.

$3°)$ Start with an ε^2-minimizer of f and choose $\lambda = \varepsilon$; then apply the results of question $2°)$.

221. *Hints.* Use the original definition of arctanh as the inverse of the hyperbolic tangent function (tanh), or the expression of $\operatorname{arctanh}(x)$ in terms of the logarithm function. Other idea: use differential calculus, prove that the gradient vectors of the functions of (x, y, z) appearing in the two sides of (1) are equal.

222. *Hint.* Use the basic definitions of eigenvalues and eigenvectors: $\lambda = u + iv \in \mathbb{C}, z = x + iy \in \mathbb{C}^n$ (where $u \in \mathbb{R}, v \in \mathbb{R}, x \in \mathbb{R}^n, y \in \mathbb{R}^n$),

$$(A + iB)(x + iy) = (u + iv)(x + iy).$$

Then express (u, v) as a convex combination of two vectors which lie in K.

224. *Hint.* Play with φ_{e_i, e_j}, where $\{e_1, e_2, ..., e_p\}$ is the canonical basis of $\mathbb{R}^p$.

227. *Hint.* Begin by proving that $f(\mathbb{R}^n)$ is a closed interval of $\mathbb{R}$.

232. *Hints.* 1°) Integrate by parts $\int_0^1 t(1 - t)\varphi''(t)\, dt$.

4°) Consider the polynomial function $f(x, y) = xy(1 - x)(1 - y)$, precisely the one appearing under the integral sign in (3). For such a function,

$$\frac{\partial^4 f}{\partial x^2 \partial y^2}(x, y) = 4 \text{ for all } (x, y) \in [0, 1] \times [0, 1],$$

$$\int_0^1 \int_0^1 f(x, y)\, dx dy = \frac{1}{36}.$$

Hence, the inequality (1) becomes an equality: $\frac{1}{36} = \frac{1}{144} \times 4$.

233. *Hint.* Describe Σ as the graph of a function $(x, \check{y}) \in D \mapsto z = f(x, y) = \sqrt{a^2 - (x^2 + y^2)}$, and observe that the term to be integrated over D is

$$\sqrt{1 + \left(\frac{\partial f}{\partial x}\right)^2 + \left(\frac{\partial f}{\partial y}\right)^2} = \frac{a}{f(x, y)}.$$

234. *Hint.* 2°) Due to the form of the solid S, a good idea is to make use of spherical coordinates (ρ, φ, θ); the solid S is then described as: $0 \leqslant \rho \leqslant 3, 0 \leqslant \varphi \leqslant \frac{\pi}{4}, 0 \leqslant \theta \leqslant 2\pi$.

The integral to be calculated is

$$vol(S) = \iiint_S dV = \int_0^{2\pi} \int_0^{\pi/4} \int_0^3 (\rho^2 \sin \varphi)\, d\rho d\varphi d\theta.$$

111

236. *Hints.* To visualize and sketch Σ as well as S, it is helpful to determine the curves obtained by intersecting the two cylinders; they are ellipses lying in the planes with equations $y = z$ and $y = -z$.

The roof of S_0 is Σ_0; its lateral surface is of the same kind.

238. *Hint.* $2°$) The form of the domain of integration and the shape of the function to be integrated suggest the changes of variables that it would be appropriate to perform.

240. *Hint.* Beware that, in S, the cylindric part is of radius 3, while the spherical part is of radius 5.

242. *Hints.* $1°$)–$3°$). To sketch the arcs Γ_α is helpful. The larger $\alpha \geqslant 2$ is, the more "curved inwards" the arc Γ_α is. Examples: Γ_1 is a quarter of the unit circle; Γ_2 is the line-segment joining the points $(1,0)$ and $(0,1)$.

An example of a Cartesian equation is, for Γ_4,

$$\sqrt{x} + \sqrt{y} = 1.$$

$4°$) Rotate the initial orthonormal basis $(O; \vec{i}, \vec{j})$ to a new one $(O; \vec{u}, \vec{v})$, with $\vec{u} = \frac{\sqrt{2}}{2}(\vec{i} - \vec{j})$ and $\vec{v} = \frac{\sqrt{2}}{2}(\vec{i} + \vec{j})$.

244. *Hint.* Due to the shape of S, it is advisable to conduct the following change of variables: $x = u^2 v,\ y = uv^2$.

246. *Hints.* Write $\|\vec{s}\|^2 = \|\vec{u}\|^2 + \|\vec{v}\|^2 + 2\|\vec{u}\|\,\|\vec{v}\|\cos(\angle(\vec{u}, \vec{v})) + \dots$. But, since $\angle(\vec{u}, \vec{v})$ and $\angle(\overline{CB}, \overline{CA}) = \widehat{C}$ add up to π, we have $\cos(\angle(\vec{u}, \vec{v})) = -\cos\widehat{C}$. Thus,

$$\|\vec{s}\|^2 = a^2 + b^2 - 2ab\cos\widehat{C} + \dots$$

Now, use the so-called law of cosines in a triangle, or theorem of AL-KASHI, which states that $c^2 = a^2 + b^2 - 2ab\cos\widehat{C}$, and so on. Whence one obtains $\|\vec{s}\|^2 = 0$.

249. *Hint.* $2°$) (a). Consider a new orthonormal basis $(O; \vec{u}, \vec{v}, \vec{w})$ built up from the points A_1, A_2, A_3 as follows:

$$\vec{u} = \frac{\overrightarrow{OA_1}}{\left\|\overrightarrow{OA_1}\right\|},\ \vec{v} = \frac{\overrightarrow{OA_2}}{\left\|\overrightarrow{OA_2}\right\|},\ \vec{w} = \frac{\overrightarrow{OA_3}}{\left\|\overrightarrow{OA_3}\right\|}.$$

Then express the (matrix of the) quadratic form $q(x, y, z) = \frac{x^2}{a^2} + \frac{y^2}{b^2} + \frac{z^2}{c^2}$ in the new basis $(O; \vec{u}, \vec{v}, \vec{w})$.

112

250. *Hint.* Several proofs, including elementary ones, are possible. Use, for example, properties of various rectangular triangles in the tetrahedron, or vector calculus (the area of the hypotenuse-face is $\frac{1}{2}\left\|\overrightarrow{AB}\wedge\overrightarrow{AC}\right\|$), or pass via the calculations of the volume of $OABC$ and of the distance from O to the plane passing through the points A, B, C.

251. *Hints.* 2°) In the calculations and answers, one has to distinguish a_0, a_{2p} (terms of even index, with $p \geqslant 1$), a_{2p+1} (terms of odd index, with $p \geqslant 0$), b_1, b_n (for $n \geqslant 2$).

Trigonometric formulas which are useful here:

$$
\begin{aligned}
\cos a \cos b &= \frac{1}{2}[\cos(a-b)+\cos(a+b)] \\
\sin a \sin b &= \frac{1}{2}[\cos(a-b)-\cos(a+b)] \\
\sin a \cos b &= \frac{1}{2}[\sin(a+b)+\sin(a-b)] \\
\cos(2a) &= 2\cos^2 a - 1 = 1 - 2\sin^2 a.
\end{aligned}
$$

252. *Hint.* 2°) The following results are useful:

- An antiderivative of the function $x^2\cos(nx)$ on $\mathbb{R}$

is $\dfrac{x^2\sin(nx)}{n} + 2\dfrac{x\cos(nx)}{n^2} - 2\dfrac{\sin(nx)}{n^3}$;

- An antiderivative of the function $x\cos(nx)$ on $\mathbb{R}$

is $\dfrac{x\sin(nx)}{n} + \dfrac{\cos(nx)}{n^2}$.

253. *Hint.* 2°) In the calculations and answers, one has to distinguish a_0, a_{2p} (terms of even index, with $p \geqslant 1$), a_{2p+1} (terms of odd index, with $p \geqslant 0$), b_n (for $n \geqslant 1$).

254. *Hints.* 2°) In the calculations and answers, one has to distinguish a_0, a_{2p} (terms of even index, with $p \geqslant 1$), a_{2p+1} (terms of odd index, with $p \geqslant 0$), b_n (for $n \geqslant 1$).

4°) Consider $x = 0$ (or π) and $x = \frac{\pi}{4}$.

Note that, for integers p, $(-1)^{\lfloor (p+1)/2 \rfloor}$ is periodic of period 4; the first values are $1, -1, -1, 1, 1, -1, -1, 1$, etc.

257. *Hint.* With $t = \tan(\frac{\theta}{2})$, we have $y = x\tan(\theta) = \frac{2tx}{1-t^2}$. Then, use the fact that t, θ and y are of the same sign.

260. *Hint.* 3°) Starting from the expression of $E'(t)$, change $\frac{\partial u}{\partial t}(t, x)$ into $\frac{\partial^2 u}{\partial x^2}(t, x)$, and conduct an integration by parts.

113

261. *Hint.* 2°) A family of solutions of the partial differential equation $\frac{\partial^2 u}{\partial t^2} = c^2 \frac{\partial^2 u}{\partial x^2}$ is provided by
$$u(x,t) = f_1(x + ct) + f_2(x - ct),$$

where f_1 and f_2 are twice continuously differentiable functions of a real variable.

262. *Hint.* 1°) There are several ways of calculating $\mathcal{A}(a,b)$ exactly: with a simple integral $2 \int_{-a}^{a} \frac{b}{a} \sqrt{a^2 - x^2}\, dx$, via a curvilinear integral $\frac{1}{2} \oint_{\mathcal{E}} (x\, dy - y\, dx)$, by performing a change of variables in $\iint_C dx dy$, and so on.

263. *Hints.* - The sought locus Γ of the center of the ellipse $\mathcal{E}$ is a piece of a simple curve in the plane. To support intuition, start with the two limiting cases: when $a = b$, in which case $\mathcal{E}$ is a circle and Γ is reduced to a point (a, a); when $b = 0$, in which case $\mathcal{E}$ degenerates to a line-segment and Γ is a quarter circle centered at O and of radius a.

- Calculate the distance $O\Omega$ from the origin O of the axes to the center Ω of the ellipse.

264. *Hint.* Use FOURIER series techniques and results

265. *Hints.* Firstly, check that f takes nonnegative values and satisfies
$$f(x) \geq \left[f(x^{1/n})\right]^n \text{ for all positive integer } n.$$

Then, develop f around the point 1, on the right with the help of the right-derivative p of f at 1, and on the left with the help of the left-derivative q of f at 1.

Additional work consists in comparing $f(x)$ with x^q on $[0, 1]$ and with x^p on $[1, +\infty)$.

266. *Hint.* Prove (1) by induction on *card* $\mathcal{F}$. At some point of the proof, the inequality $\left(\sqrt{ab} + \sqrt{cd}\right)^2 \leq (a + c)(b + d)$ for positive a, b, c, d is helpful.

267. *Hint.* One can stay stuck for a long time if one does not deal with mathematical calculations in the right way... Set $a = \sqrt{11 + 2\sqrt{29}}$ and $b = \sqrt{11 - 2\sqrt{29}}$; then calculate $a + 2b\sqrt{5} + 5 + b^2$ and $(a + b)^2$.

268. *Hints.* 1°) $S + T$ and U are integers with 44 digits; emphasize the first 10.

2°) ε_r is very small, suggesting that $S + T$ and U are "numerically equal".

269. *Hints.* Perform calculations in the Euclidean (hence factorial) ring $\mathbb{Z}\left[i\sqrt{2}\right]$, in order to prove that:

- the equation $y^2 = x^3 - 2$ has infinitely many rational solutions but only one solution involving positive integers, $(x, y) = (3, 5)$;

- the equation $y^2 = x^3 + 2$ has no solution involving positive integers x and y.

114

275. *Hint.* A key-ingredient is to study the sequence $(\|x_k - y\|)_k$, where y is arbitrarily chosen in S. Such a sequence is decreasing, and the sequence $(x_k)_k$ is bounded.

278. *Hint.* The solution of $(\mathcal{C}_3)$ is defined on the whole $\mathbb{R}$. To get it, "glue together" the solutions of $(\mathcal{C}_1)$ and $(\mathcal{C}_2)$ when they cross the Ox axis.

A discussion on the values of b is necessary; three situations have to be considered: when $|b| \leqslant 1$, when $b > 1$, when $b < -1$.

280. *Hint.* A proof by contradiction is suggested. Use simple techniques and results on differentiable functions of a real variable.

281. *Hint.* Decompose $P(x)$, and also $P'(x)$, as products of monomials $(x - \alpha_i)^{m_i}$, and study the behavior of $\frac{P'(x)}{P(x)}$ as x approaches α_i.

284. *Hint.* One can try to answer by considering progressively larger and larger classes of functions: differentiable functions (the simplest case!), locally LIPSCHITZ or absolutely continuous functions, etc., but the question is posed for functions enjoying no further properties than the one posed at the beginning.

285. *Hint.* The result is known for $g(x) = \sin(\pi x)$ and $g(x) = \cos(\pi x)$. This is the so-called LEBESGUE–RIEMANN lemma (see Tapa 49). It therefore holds true for the functions $g(x) = \sin(k\pi x)$, $g(x) = \cos(k\pi x)$, and

$$g(x) = \sum_{k=1}^{n} \left[a_k \cos(k\pi x) + b_k \sin(k\pi x) \right]. \tag{1}$$

Then, any continuous and 1-periodic function g can be uniformly approximated by functions like the ones displayed in (1).

286. *Hint.* Prove the nonnegativity of Δ_n on $\mathbb{R}$ by induction on n. Key: if r is a real root of P, then $P(x) = (x - r)Q(x)$, where Q now has $n - 1$ real roots.

287. *Hint.* Begin by proving that f is of class $\mathcal{C}^\infty$. Then derive with respect to x and with respect to y the functional relations (1) and (2).

288. *Hint.* Since the Hessian (matrix) $\nabla^2 f$ of f is a continuous mapping from $\mathbb{R}^n$ into $\mathcal{S}_n(\mathbb{R})$, start with local "diff-convex" decompositions of f around x, $f = r \|\cdot\|^2 - (r \|\cdot\|^2 - f)$, and then "glue" them.

289. *Hint.* 1°) Try affine functions $x \mapsto f(x) = ax + b$.

290. *Hint.* Begin by considering uniformly continuous functions f.

293. *Hints.* 1°) - We have $s = u + v$, $t = u - v$. Use trigonometric formulas to obtain (1).

- Since $v \in (0, \pi]$, we have $0 \leqslant \frac{\sin v}{v} < 1$. Derive from the expression (1) that $\|q\| < 1$.

2°) The set Q is a bizarre non-convex subset contained in the open unit ball $B(0,1)$ in $\mathbb{R}^2$; it is shaped like the complementary set in $B(0,1)$ of a cardioid; hence the intersection of Q and $f'([0, 2\pi])$ is empty. However, by close-convexifying it, we obtain the closed unit ball of $\mathbb{R}^2$.

295. *Hint.* Let $M = \displaystyle\sum_{i=1}^{d} M_i$. Calculate $M_i M$ for all i, show that M is nilpotent (with the help of the trace), calculate M^k.

296. *Hint.* Use the characteristic polynomial or the minimal polynomial of M.

298. *Hint.* To answer the second question, use the fact (to be proved) that any nilpotent matrix is similar to a matrix with zeros on the diagonal.

304. *Hints.* 1°) Relation (1) is a consequence of the classical CAUCHY–SCHWARZ inequality in $\mathbb{R}^n$ by observing that:

$$|x^T y| \;\leqslant\; \sum_{i=1}^{n} |x_i| \times |y_i| = \sum_{i=1}^{n} \max\left(|x_i|, |y_i|\right) \times \min\left(|x_i|, |y_i|\right);$$
$$(\max\left(|x_i|, |y_i|\right))^2 \;=\; \max(x_i^2, y_i^2), \; (\min\left(|x_i|, |y_i|\right))^2 = \min(x_i^2, y_i^2).$$

2°) As above, relation (3) is a consequence of the classical CAUCHY–SCHWARZ inequality in $\mathcal{C}\left([a, b], \mathbb{R}\right)$ by observing that:

$$\left| \int_a^b f(x) g(x) \, dx \right| \;\leqslant\; \int_a^b |f(x) g(x)| \; dx = \int_a^b \max(|f(x)|, |g(x)|) \times \min(|f(x)|, |g(x)|) \, dx;$$
$$(\max(|f(x)|, |g(x)|))^2 \;=\; \max(f^2(x), g^2(x)), \; (\min(|f(x)|, |g(x)|))^2 = \min(f^2(x), g^2(x)).$$

305. *Hints.* - The property (1) implies that A is positive semidefinite.

- Begin with a positive definite A. Consider $a \neq 0, b \neq 0$ and the positive real number $\alpha = \left(\frac{a^T A a}{b^T A b}\right)^{1/4}$.

Then use the following expansions and their squares:

$$a^T M b \;=\; \frac{1}{4}\left[(a + b)^T M(a + b) - (a - b)^T M(a - b)\right];$$
$$a^T M b \;=\; \left(\frac{a}{\alpha}\right)^T M(\alpha b).$$

306. *Hint.* Clearly, positive definite A and negative definite A satisfy (1) and (2). But there are others... To determine them, use the spectral decomposition of A (and thus that of A^{-1}).

307. *Hints.* We have:

$$\dim(\ker \varphi) + \dim(\operatorname{Im} \varphi) = \dim(\mathcal{S}_n(\mathbb{R})) = \frac{n(n+1)}{2};$$

$$\dim(\operatorname{Im} \varphi) \leqslant \dim(\mathcal{AS}_n(\mathbb{R})) = \frac{n(n-1)}{2}.$$

Observe that (1) is equivalent to $\varphi(S) = 0$.

308. *Hints.* Inequality (1) is just the "homogenized" form of inequality (2). For the implication $[(2) \Rightarrow (1)]$, use tA and tB for $t > 0$, factor out t^4 and let t tend to 0.

Inequality (3) is a rewriting of (1) using only traces. To prove (3), one can make use of the symmetric bilinear form b on $\mathcal{M}_2(\mathbb{R})$ defined by $b(A, B) = \operatorname{tr}(AB) - \frac{(\operatorname{tr}A)\times(\operatorname{tr}B)}{2}$, which is nonnegative when restricted to the subspace of symmetric matrices.

309. *Hints.* The proposed equivalences are tougher to establish than what one might think at first glance... $[(1) \Leftrightarrow (2)]$ (sometimes called OGAWA's theorem) is more involved than $[(1) \Leftrightarrow (3)]$ (sometimes called the CRAIG–SAKAMOTO theorem). Things would have been easier under the hypothesis that A and B commute... but this is not assumed (that will be a consequence of the equivalences).

One can use resources either from Linear algebra (properties of traces and determinants of matrices) or from Analysis (expanding $\det(I_n + zA)$ as the sum of power series (see Tapa 202), or differential calculus).

311. *Hints.* 1°) Since both $(1 - t)I_n + tS_1$ and S_2 are positive definite,

$$d(t) = \det M(t) = \det[(1 - t)I_n + tS_1] \times \det S_2 > 0 \text{ for all } 0 \leqslant t \leqslant 1.$$

Hence the symmetric matrix $M(t)$ is always nonsingular for $0 \leqslant t \leqslant 1$.

2°) The "variational" expression of $\lambda(t)$ as the minimum of the quadratic form $x^T M(t)x$ over the set of unit vectors x of $\mathbb{R}^n$ allows us to show easily that λ is a continuous function. Now, the continuous function λ goes from $\lambda(0) =$ smallest eigenvalue of $M(0) = S_2$ to $\lambda(1) =$ smallest eigenvalue of $M(1) = S_1 S_2$. It never crosses 0 because $M(t)$ is always nonsingular for $0 \leqslant t \leqslant 1$. Hence $\lambda(1)$ is of the same sign as $\lambda(0)$, that is to say: $\lambda(1) > 0$. Consequently, $S_1 S_2$ is positive definite.

313. *Hint.* The n nonnegative real numbers $\sigma_i(M)$ are called the *singular values* of M. The number of non-null singular values of M is exactly the rank of M. The key-tool here is the so-called singular value decomposition of matrices: for M in $\mathcal{M}_n(\mathbb{R})$, there exist orthogonal matrices U and V such that

$$M = U\Sigma V^T, \text{ where } \Sigma = \operatorname{diag}(\sigma_1(M), \sigma_2(M), ..., \sigma_n(M)).$$

315. *Hint.* Begin by proving the results for a square.

316. *Hint.* 1°) Proof by induction on n.

324. *Hint.* 3°) Prove it by contradiction.

325. *Hints.* The answer is Yes and this is somehow surprising. Here are hints for a proof. Since the function f is positively homogeneous (*i.e.*, $f(\lambda A, \lambda x) = \lambda f(A, x)$ for all $\lambda > 0$ and (A, x) in $\mathcal{P}_n \times \mathbb{R}^n$), it suffices, to ensure the convexity of f, to prove its subadditivity, *i.e.*,

$$f(A_0 + A, x_0 + x) \leqslant f(A_0, x_0) + f(A, x)$$

for (A_0, x) and (A, x) in $\mathcal{P}_n \times \mathbb{R}^n$. Consider therefore

$$I = x_0^T A_0^{-1} x_0 + x^T A^{-1} x - (x_0 + x)^T (A_0 + A)^{-1} (x_0 + x).$$

We intend to prove that $I \geqslant 0$.

Then: decompose A_0 as CC^T, where C is an invertible matrix; there exists a diagonal matrix D with positive diagonal terms d_i and an orthogonal matrix U such that $A = (CU) D (U^T C^T)$; change to new variables $y = (CU)^{-1} x$ and $y_0 = (CU)^{-1} x_0$; reformulate I and conclude that $I \geqslant 0$.

326. *Hints.* This is an excellent opportunity to use integral formulas on surfaces of revolution:

- The volume of a solid of revolution obtained by rotating the smooth curve $y = f(x) > 0$ for $x = a$ to $x = b$ around the x-axis is $\mathcal{V} = \pi \int_a^b [f(x)]^2 \, dx$.

- The area of the surface of revolution obtained by rotating the smooth curve $y = f(x) > 0$ for $x = a$ to $x = b$ around the x-axis is $\mathcal{A} = 2\pi \int_a^b f(x)\sqrt{1 + [f'(x)]^2} \, dx$.

2°) Beware, the result is counterintuitive.

329. *Hint.* The four vectors are $\vec{v_1} = \frac{1}{2}\overrightarrow{BC} \wedge \overrightarrow{BA}$, $\vec{v_2} = \frac{1}{2}\overrightarrow{BA} \wedge \overrightarrow{BD}$, $\vec{v_3} = \frac{1}{2}\overrightarrow{BD} \wedge \overrightarrow{BC}$ and $\vec{v_4} = \frac{1}{2}\overrightarrow{DA} \wedge \overrightarrow{DC}$. Indeed, the lengths of these vectors are equal to the areas of the corresponding faces; the vector-product of two vectors is orthogonal to the plane determined by these vectors, and it points outward because of the right-hand rule. It remains to prove that the four vectors $\vec{v_i}$ add up to the zero vector.

331. *Hints.* 1°) Beware, here and for the calculations coming afterwards, that $f(x)$ is not e^x for x between -2π and 0.

2°) The following results are useful:

an antiderivative of $e^x \cos(nx)$ on $\mathbb{R}$ is $\dfrac{e^x}{n^2 + 1} \left[\cos(nx) + n\sin(nx)\right]$;

an antiderivative of $e^x \sin(nx)$ on $\mathbb{R}$ is $\dfrac{e^x}{n^2 + 1} \left[\sin(nx) - n\cos(nx)\right]$.

4°) Choose successively $x = 1$ and $x = -1$.

5°) Two ways of obtaining this sum: either choose $x = 0$ in (1), or use PARSEVAL's theorem.

332. *Hints.* As is often the case in this context, use FOURIER series techniques and results. The function f is extended to an odd function on $[-\pi, \pi]$ and then to a 2π-periodic function on $\mathbb{R}$. Since $\mathcal{I}_1(\sin) = 0$, what has to be proved is:

$$\int_0^\pi [f'(x)]^2 \, dx \leqslant \int_0^\pi [f(x)]^2 \, dx,$$

with equality only when $f = \sin$. Then use the relations linking FOURIER coefficients of f and f', and PARSEVAL relations.

119

Chapter 3. Answer(s), Comment(s)

1. *Comments.* 1°) - With the ordered set of real numbers $-y_n \leqslant -y_{n-1} \leqslant \ldots \leqslant -y_1$, inequality (1) leads to the companion inequality

$$\sum_{k=1}^{n} x_k y_{n-k+1} \leqslant \sum_{k=1}^{n} x_k y_{\sigma(k)}. \tag{1'}$$

   - Inequality (1) is called the *rearrangement inequality.* Note that no assumptions were made concerning the sign of the real numbers $x_1 \leqslant x_2 \leqslant \ldots \leqslant x_n$ and $y_1 \leqslant y_2 \leqslant \ldots \leqslant y_n$.

   2°) - Inequality (2) is called TCHEBYCHEV's *sum inequality.*

   - As expected, there is a "continuous" form of inequality (1): If $f, g : [a, b] \to \mathbb{R}$ are two continuous increasing functions, then

$$\left(\frac{1}{b-a} \int_a^b f(x)\, dx\right) \times \left(\frac{1}{b-a} \int_a^b g(x)\, dx\right) \leqslant \frac{1}{b-a} \int_a^b f(x)g(x)\, dx. \tag{4}$$

3. *Answer.* The lower bound n^2 is attained for $x_1 = x_2 \ldots = x_n = \frac{1}{n}$.

4. *Answer.* 2°) The minimal value is achieved when $x_i = \frac{1}{a_i}\mu$ for all $i = 1, 2, \ldots, n$.

5. *Answer.* The identity permutation is the only maximizer.

6. *Answers.* 2°) When $x = y = 1/2$.

   3°) We have:

$$\sum_{i=1}^{n} x_i \log(x_i) \geqslant -\log(n). \tag{2}$$

8. *Answers.* $S_o = S_e = 2^{n-1}$.

9. *Answer.* 1°) We have $S_n + S_n = [1 + (2n-1)] + \ldots + [(2n-1) + 1] = n(2n)$. Hence, $S_n = n^2$.

   Comment. The property on successive odd integers proved in question 2°) goes back to GALILEO (1615).

10. *Answer.* The limit in (1) is $\frac{1}{4}$.

    Comment. A more general result would be:

$$\text{For any positive integer } k, \quad \lim_{n \to +\infty} \frac{1^k + 2^k + \ldots + n^k}{n^{k+1}} = \frac{1}{k+1}.$$

© Springer International Publishing Switzerland 2016
J.-B. Hiriart-Urruty, *Mathematical Tapas*, Springer Undergraduate
Mathematics Series, DOI 10.1007/978-3-319-42186-5_3

11. *Answers.* 1°) $S_n(1) = \frac{n(n+1)}{2}$, $S_n(2) = \frac{n(n+1)(2n+1)}{6}$, $S_n(3) = \frac{n^2(n+1)^2}{4} = [S_n(1)]^2$.

3°) $S_n(5) = \frac{1}{12}n^2(n+1)^2(2n^2+2n-1)$; $S_n(7) = \frac{1}{24}n^2(n+1)^2(3n^4+6n^3-n^2-4n+2)$.

14. *Answer.* Let d be the distance (in miles) between the home of the student and the university. The duration of the trip to the university is (in hours) $T_1 = \frac{d}{10}$. Similarly, if $v_{back} > 0$ is the average velocity on the way back, the duration of the trip back is $T_2 = \frac{d}{v_{back}}$. The average velocity v_{aver} on the journey (round-trip) is therefore

$$v_{aver} = \frac{2d}{T_1 + T_2} = \frac{2}{\frac{1}{10\ m/h} + \frac{1}{v_{back}}}. \tag{1}$$

For example, if we take $v_{back} = 30$ m/h, the resulting v_{aver} is 15 m/h, and not 20 m/h.

So, how to get $v_{aver} = 20$ as asked for? The answer is: it is impossible! One cannot have $v_{aver} = 20$ in (1).

15. *Comment.* This is a classical result. Proving it is very instructive.

16. *Answers.* The largest element in $\{f(n)\,|\,n$ positive integer$\}$ is $f(3) = \sqrt[3]{3}$.

   Comment. An application. If $m > n \geqslant 1$, we have $\sqrt[n]{n} > \sqrt[m]{n}$.

19. *Answer.* Never, except for $n = 0$ or $m = 0$.

   To see why, observe that $\binom{n+m}{n} = \binom{n+m}{m} = \frac{(n+m)!}{n!m!}$.

21. *Answer.* 2°) $S_p = 1 + \frac{1}{p}$.

22. *Answer.* 1°) Yes. It suffices to consider the functions of the form $f(x) = rx$ with an irrational coefficient r, for example $r = \sqrt{2}$.

   2°) The curve $\mathcal{C}_2$ contains points with rational coordinates, $(\frac{3}{5}, \frac{4}{5})$ for example. For $n \geqslant 3$, according to the FERMAT–WILES theorem (1995), the set described in (2) is empty; hence, the curve $\mathcal{C}_n$ avoids all the points with positive rational coordinates.

27. *Answer.* Every real number is a limit point of a sequence of elements in S. Hence, the closure of S is the whole $\mathbb{R}$.

28. *Answer.* Yes, there are such functions f; an example is $f(x) = x^2 \sin(x)$.

30. *Answer.* They are constant.

32. *Comment.* We add to the first result the following one: $(m_p)^p = \sum_{i=1}^{n} x_i^p$ (not m_p) converges to $card\{i\,|\,x_i \neq 0\}$ as $p \to 0$.

35. *Answer.* Yes, both f and g are necessarily continuous.

36. *Answer.* Just check that formula (1) holds with the help of:

$$\sin(a \pm b) = \sin(a)\cos(b) \pm \sin(b)\cos(a);$$
$$\cos^2(x) = -\sin^2(x) + 1.$$

37. *Comment.* (2) is known as MECHAIN's formula.

38. *Answer.* 2°) The constant value is $\frac{\pi}{4}$.

40. *Comments.* - The function g is called LAMBERT's function; it is not possible to express it explicitly in terms of the usual functions (polynomial, exponential, logarithm,...). It plays a role in several areas of mathematics and physics.

- A step further and we obtain the expression for $g''(y)$; we note that g is actually a concave function.

- Another example of a differential equation where a solution can be expressed in terms of g is the following one: $x < 0 \mapsto \varphi(x) = -g(-x)/x$ solves the differential equation $[1 - xy(x)]\,y'(x) = y(x)^2$.

41. *Answers.* 1°) All five derivatives equal $\frac{1}{\cos x}$. Since

$$f_1(0) = f_2(0) = f_3(0) = f_4(0) = f_5(0) = 0,$$

the five functions are equal. In short,

$$f(x) = \int_0^x \frac{dt}{\cos t}. \tag{2}$$

2°) f is a strictly increasing differentiable bijection from I onto $\mathbb{R}$. We get from (1-1) and (1-2) four different expressions for $g(x)$:

$$g(x) = \arctan(\sinh x); \ g(x) = \arcsin(\tanh x); \tag{3-1}$$
$$g(x) = 2\arctan(\tanh\frac{x}{2}); \ g(x) = 2\arctan(e^x) - \frac{\pi}{2}. \tag{3-2}$$

The derivative of g is $g'(x) = \frac{1}{\cosh x}$. Since $g(0) = 0$, we have

$$g(x) = \int_0^x \frac{dt}{\cosh t}. \tag{4}$$

Putting together formulas (2) and (4), we have the following beautiful "symmetric" relationship:

$$\left(x \in \left(-\frac{\pi}{2}, \frac{\pi}{2}\right) \text{ and } y = \int_0^x \frac{dt}{\cos t}\right) \Leftrightarrow \left(y \in \mathbb{R} \text{ and } x = \int_0^y \frac{dt}{\cosh t}\right). \tag{5}$$

Comments. - The function f, whose four different expressions have been given in (3-1) and (3-2) (but there are others) is called GUDERMANN's function. GUDERMANN (1798–1851) was WEIERSTRASS's favorite teacher. In a certain sense, formula (5) relates cos and cosh without passing through the complex numbers.

- For more on some "particular" functions, see

J.-B. HIRIART-URRUTY, *Des fonctions... pas si particulières que ça : celles de* LAMBERT, GUDERMANN *et* AIRY. To appear in Quadrature (2016).

42. *Answer.* Second part of 2°) : No; the function displayed in the first question does satisfy (1), but it is not uniformly continuous on $\mathbb{R}$.

Comment. Beware, the product of two uniformly continuous functions is not uniformly continuous, for example: $f(x) = x \times x$.

43. *Comments.* - If we collect all the limit points

$$\{l : f'(x_k) \to l, \text{ with } f \text{ differentiable at all } x_k \text{ and } x_k \to 0\}, \tag{2}$$

we get here the whole interval $[-1, 1]$. This line-segment, containing $f'(0) = 0$, represents what should be a "generalized derivative" of f at 0; it takes into account the behavior of f around 0.

- For a locally LIPSCHITZ $f : \mathbb{R} \to \mathbb{R}$, the line-segment defined as the smallest interval containing

$$\{l : f'(x_k) \to l, \text{ with } f \text{ differentiable at all } x_k \text{ and } x_k \to x\} \tag{3}$$

is the so-called CLARKE's generalized derivative at x (1973); it is denoted $\partial f(x)$.

The following "tilted" versions of f in (1) are interesting to consider with respect to the inversion of f around 0 :

$$\begin{aligned} g(x) &= f(x) + \frac{1}{2}x, \\ h(x) &= f(x) + 2x. \end{aligned}$$

We have $g'(0) \neq 0, h'(0) \neq 0$ but neither g nor h is continuously differentiable. Hence, the usual inverse function theorem does not apply at 0. We observe that g is not monotone in any neighborhood of 0 (and $0 \in \partial g(0) = \left[-\frac{1}{2}, \frac{3}{2}\right]$), while h can be inverted in a neighborhood of 0 (the reason is that, with the factor $2x$, the graph of the function f has been "stretched" enough; indeed $0 \notin \partial h(0) = [1, 3]$). There is an inverse function theorem for locally LIPSCHITZ functions behind this.

44. *Answer.* The convergence result (1) may be surprising since just the differentiability of f at a is assumed.

If the sequences (x_n) and (y_n) are not "straddling" the point a, anything may happen: non-convergence of the slope $\frac{f(y_n)-f(x_n)}{y_n-x_n}$, convergence towards a limit which is not $f'(a)$, etc. If f is now assumed to be continuously differentiable at a, then the slope $\frac{f(y_n)-f(x_n)}{y_n-x_n}$ necessarily converges to $f'(a)$.

45. *Answer.* The function $\theta : x \mapsto f(x) - xf'(c)$ is injective and continuous, hence strictly monotone. Thus, the derivative $\theta' : x \mapsto f'(x) - f'(c)$ is of a constant sign, and c is an extremum point of f'. Consequently the derivative of f' at c, $f''(c)$, should be equal to 0.

46. *Answer.* The announced limit is $-\frac{f''(a)}{2[f'(a)]^2}$.

49. *Comments.* - One similarly obtains (1-2) by considering $a_n = \frac{2}{\pi} \int_0^\pi f(x) \cos(nx)\, dx$ and S_n modified accordingly.

The results (1-1) and (1-2) are known as the LEBESGUE–RIEMANN lemma.

- As a general rule, "oscillating forces the integrals to 0". Let $f : \mathbb{R} \to \mathbb{R}$ be a continuous function; then, for any a and b :

$$\int_a^b f(x) \cos(nx)dx \to 0 \text{ as } n \to +\infty. \tag{2}$$

50. *Comment.* It is easy to illustrate the result graphically.

53. *Answer.* 2°) Such a situation is possible, here is an example: $f(x) = x$ for all $x \in \left[0, \frac{1}{2}\right]$, $f(x) = 1 - x$ for all $x \in \left[\frac{1}{2}, 1\right]$.

60. *Answer.* 2°) One has:

$$\frac{1}{n!} f^{(n)}\left(\frac{1}{n}\right) = 1 + \frac{1}{2} + \ldots \frac{1}{n} - \log(n),$$

whose limit value, as $n \to +\infty$, is the so-called EULER–MASCHERONI constant.

63. *Answer.* 2°) The solution of $(\mathcal{C}_2)$ is 2π-periodic. The solution of $(\mathcal{C}_3)$, with $x_0 \neq 0$, is not 2π-periodic.

65. *Answer.* $L_n \to \frac{4}{\pi}$ as $n \to +\infty$.

66. *Answer.* $L = n$.

67. *Answer.* The solutions are $1, 2, -2$, up to a permutation of these three numbers.

69. *Answer.* It is indeed enough to have (2). Using determinants, a consequence of (2) is that $\det(A)\det(B) = \det(AB) = \det(I_n) = 1 \neq 0$; whence $\det(A) \neq 0$ and A is invertible.

Comment. There is a result from Linear algebra behind this: a linear mapping from a finite-dimensional vector space to itself is bijective if and only if it is injective, if and only if it is surjective; thus, the right invertibility implies invertibility.

70. *Answer.* The expectation of $\det M$ is 0.

71. *Answers.* 1°) $A^{-1} = A^*$. Furthermore, $A^p = I_2$ if p is even, $A^p = A$ if p is odd.

2°) For example, $U = \frac{1}{\sqrt{2}} \begin{bmatrix} 1 & 1 \\ -i & i \end{bmatrix}$.

72. *Comments.* - A "matricial" version of the classical inequality (3) is as follows. Let $(a_{i,j})_{\substack{1 \leqslant i \leqslant n \\ 1 \leqslant j \leqslant m}}$ be a collection of real numbers. Then

$$\sum_{i=1}^{n} \sum_{j=1}^{m} a_{i,j} \leqslant \sqrt{n \sum_{i=1}^{n} \sum_{j,k=1}^{m} a_{i,j} a_{i,k}}. \tag{4}$$

To get at this, it suffices to apply (3) with $x_i = 1$ and $y_i = \sum_{j=1}^{m} a_{i,j}$ for all $i = 1, 2, ..., n$.

- Besides the classical case with two collections of real numbers, one could imagine a CAUCHY–SCHWARZ inequality with three collections. For $x = (x_1, ..., x_n), y = (y_1, ..., y_n), z = (z_1, ..., z_n)$, the targeted inequality is:

$$\left(\sum_{i=1}^{n} x_i y_i z_i \right)^2 \leqslant \|x\|^2 \times \|y\|^2 \times \|z\|^2. \tag{5}$$

Actually, this is nothing new: putting $z_i' = z_i / \|z\|$, we have $|z_i'| \leqslant 1$ for all i, so that

$$\sum_{i=1}^{n} |x_i y_i z_i'| \leqslant \sum_{i=1}^{n} |x_i y_i|$$
$$\leqslant \|x\| \times \|y\| \quad \text{[from the classical inequality (3)]}.$$

Whence (5) follows.

76. *Comment.* (1) is called the KY FAN inequality.

77. *Comments.* - A better result than the one derived in 2°) (on the compared behaviors of H_n and $\ln(n)$ as n goes to $+\infty$) can be obtained; indeed, there exists a constant $\gamma \approx 0.5772$ (called the EULER–MASCHERONI constant) such that

$$H_n - \ln(n) = \gamma + \frac{1}{2n} - \frac{1}{12n^2} + \frac{1}{120n^4} + O\left(\frac{1}{n^6}\right). \tag{2}$$

- The convergence of $a_n = H_n - \ln(n)$ towards γ is slow; it can be improved as follows. Clearly, γ is also the limit of $b_n = H_{n-1} - \ln(n)$. Thus, $c_n = \frac{a_n + b_n}{2}$ converges towards γ as well, but faster than a_n or b_n. This "numerical observation" can be explained by a better polyhedral approximation in $c_n - \gamma$ than in $a_n - \gamma$, of appropriate underlying integrals $\int_{n-1}^{n} \frac{1}{t}\, dt$.

78. *Answer.* The limit of $S_p(n)$, as n goes to $+\infty$, is $\ln(p)$.

79. *Answers.* $1°)$ (a) $k_1 = 1, k_2 = 4, k_3 = 11$.

(b) The sequence $(k_n)_n$ is integer-valued, increasing, and goes to $+\infty$ as $n \to +\infty$.

$2°)$ $\frac{k_{n+1}}{k_n} \to e$ as $n \to +\infty$.

Comments. - A nice formula mixing series with general terms $\frac{1}{n}, \frac{1}{n^2}$ and $\frac{1}{n^3}$ is the following, due to EULER:

$$\sum_{n=1}^{+\infty} \frac{1}{n^3} = \sum_{n=1}^{+\infty} \frac{H_n}{n^2}. \tag{2}$$

- As a general rule, some sums $S_k = \sum_{n=1}^{+\infty} \frac{1}{n^k}$ (with an integer $k \geqslant 2$) are easy to obtain, some not. For example, with the help of FOURIER series, we find:

$$S_2 = \frac{\pi^2}{6}, \quad S_4 = \frac{\pi^4}{90}. \tag{3}$$

For S_3: there is no closed-form expression like in (3); it has been proved by R. APÉRY (1977) that S_3 is irrational; an approximate numerical value of S_3 is 1.20.

82. *Answer.* $(x, y, z) = (3, 2, 1)$ and the five other triples obtained by permutations of the integers $3, 2, 1$.

83. *Comment.* A better result than the one derived in $2°)$ can be obtained; indeed,

$$n! = \sqrt{2\pi n} \left(\frac{n}{e}\right)^n \left(1 + \frac{1}{12n} + \frac{1}{288n^2} - \frac{139}{5140n^3} + O\left(\frac{1}{n^4}\right)\right). \tag{2}$$

A consequence is:

$$n! \sim \sqrt{2\pi n} \left(\frac{n}{e}\right)^n \quad \text{as } n \to +\infty.$$

This is called STIRLING's formula.

85. *Answer.* Answer to the first question: $\binom{n-1}{k-1}$. For the second question, use the result of the first question with $k, k+1, \dots, n$, so that the final number is: $\sum_{m=k}^{n} \binom{m-1}{k-1} = \binom{n}{k}$.

88. *Comments.* - A consequence of the result above is that the series with general term $u_n = \frac{1}{n^2}$ is convergent, with

$$\sum_{n=1}^{+\infty} \frac{1}{n^2} \leqslant 2.$$

- We know that the series with general term $\frac{1}{n} = \frac{n}{n^2}$ is divergent. A more general result is: for a sequence of integer-valued a_n, such that $a_{n'} \neq a_n$ whenever $n' \neq n$, the series with general term $\frac{a_n}{n^2}$ is divergent.

90. *Answer.* We have

$$u_n = (2n)! \left(1 + \frac{1}{2} + \dots + \frac{1}{n} \right).$$

92. *Answer.* No, even if the property (1) looks pretty much like the CAUCHY property for the sequence $(u_n)_{n \geqslant 1}$.

A counterexample is given by $u_n = \ln(n)$ or $u_n = \sum_{k=1}^{n} \frac{1}{k}$.

93. *Answers.* 1°) No, a subadditive sequence is not always a convergent one; take for example $u_n = n$ or $u_n = \sqrt{n}$.

2°) The sequence $\left(\frac{u_n}{n} \right)_{n \geqslant 1}$ converges towards $l = \inf_{n \geqslant 1} \left(\frac{u_n}{n} \right)$.

95. *Answer.* The (positive) general term $u_n = \frac{1}{1^2 + 2^2 + \dots \, n^2} = \frac{6}{n(n+1)(2n+1)}$ is equivalent to $\frac{3}{n^2}$ as $n \to +\infty$; hence the series with general term u_n is convergent.

To calculate $\sum_{n=1}^{+\infty} u_n$, first decompose u_n as $u_n = \frac{6}{n} + \frac{6}{n+1} - \frac{24}{2n+1}$. Consequently,

$$\sum_{n=1}^{+\infty} u_n = 18 - 24 \log(2).$$

96. *Answer.* $S = \sum_{k=1}^{+\infty} \frac{(-1)^{k+1}}{k} = \ln(2)$.

97. *Comments.* 1°) The sequence (a_n) is indeed convergent, but its limit is not necessarily 0.

2°) Under the assumption made for this question, the sequences (a_n) and (b_n) are not necessarily convergent.

98. *Comment.* The conditions imposed in the two situations look pretty similar, but the assumptions made on the sequences (α_n) and (β_n) are not comparable.

99. *Answers.* 1°) We have, for all $n \geqslant 1$,

$$z_n = r e^{i(\theta/2^n)} \prod_{k=1}^{n} \cos \left(\frac{\theta}{2^k} \right).$$

127

2°) If $\theta = 0$, $z_n = z_0$ for all n.

If $\theta = \pi$, $z_n = 0$ for all $n \geqslant 1$.

If $\theta \in (0, \pi)$ or if $\theta \in (-\pi, 0)$, then $\lim_{n \to +\infty} z_n = r \frac{\sin(\theta)}{\theta}$.

Comment. The formula (1) has its own interest; in an equivalent form, it reads

$$\left[\prod_{i=1}^{n-1} \cos(2^i x) \right] \times \sin(x) = \frac{\sin(2^n x)}{2^n}. \tag{1bis}$$

As an example, by taking $x = \frac{\pi}{9}$ and $n = 3$, we obtain

$$\cos\left(\frac{\pi}{9}\right) \cos\left(\frac{2\pi}{9}\right) \cos\left(\frac{4\pi}{9}\right) = 1. \tag{2}$$

This is known as MORRIE's trigonometric relation.

100. *Comment.* This property, at least its extension to sequences in $\mathbb{R}^d$, is useful in proving convergence results for algorithms designed in Optimization and Numerical analysis.

102. *Answer.* The answer is Yes. The odd function $f(x) = \frac{x}{1+|x|}$ has the required property.

104. *Answer.* 1°) Only two functions meet the requirements: $x \mapsto f(x) = x$ and $x \mapsto f(x) = 1 - x$.

105. *Answer.* The function f is periodic of period $T = 4(s - r)$.

106. *Answer.* 3°) The evoked differential equation is: $f'(x) = f(0)f(x)$.

The non-null solutions f of (1) are $x \in \mathbb{R} \mapsto f(x) = \exp(rx)$, where r is a real number.

107. *Answer.* Only the null function $f(x) = 0$ and the identity function $f(x) = x$.

108. *Answers.* 1°) $f(x) = \sin(x)$ for example.

2°) The solutions of the posed problem are the functions $x \in \mathbb{R} \mapsto f(x) = A \sin(x) + B[1 - \cos(x)]$, with A and B real numbers.

109. *Comments.* - To remember (3), it suffices to note that only cosines of sums $\varepsilon_1 a_1 + \varepsilon_2 a_2 + \ldots + \varepsilon_n a_n$ appear in the right-hand side of (3). Which ones? All of them (there are 2^n possibilities), and one then takes their mean value.

- An explanation of (3) coming from Probability theory (with BERNOULLI random variables) can be found in

J.-B. HIRIART-URRUTY, *Les formules de trigonométrie sans pleurs...* Bulletin de l'APMEP (Mathematics Teachers Association), n° 515, 407–410 (2015).

110. *Answers.* Solution of $(\mathcal{C}_1)$: $x(t) = t^2/2$;

Solution of $(\mathcal{C}_2)$: $x(t) = 2\exp(t-2)$;

Solution of $(\mathcal{C}_3)$: $x(t) = 0$ if $t \leqslant 0$; $x(t) = t^2/2$ if $0 \leqslant t \leqslant 2$; $x(t) = 2\exp(t-2)$ if $2 \leqslant t$.

111. *Answers.* 1°) $f'(x) = \frac{1-x}{\ln(x)}$.

2°) $\lim_{x \to 0} f(x) = 0$; $\lim_{x \to 1} f(x) = -\ln(2)$.

112. *Answers.* 1°) and 2°). Let $a \in I$. The function f, defined to be equal to 1 if $x \neq a$ and 0 if $x = a$ disproves 1°) and the first part of 2°). The function $f(x) = |x|$ disproves the second part of 2°).

3°) The statement is true.

113. *Answers.* 1°) $J_R = \frac{\pi}{4}(1 - e^{-R^2})$.

2°) $J_R \leqslant I_R \leqslant J_{R\sqrt{2}}$.

3°) $\lim_{R \to +\infty} I_R = \lim_{R \to +\infty} J_R = \frac{\pi}{4}$. Hence, $\int_0^{+\infty} e^{-x^2}dx = \frac{\sqrt{\pi}}{2}$.

116. *Answer.* The integral I_n tends to $f(1)$ as $n \to +\infty$.

120. *Answer.* There are no (real) solutions to this equation.

121. *Answer.* No, f is not differentiable at ± 1. The theorem of composition of differentiable functions (also called the chain rule) does not apply at $x = 1$ since the arcsin function is not differentiable at $y = 1 = \frac{2 \times 1}{1 + 1^2}$. So, to check whether f is differentiable at $x = 1$ or not, we have either to go back to the definition of the derivative of f at 1, or consider the possible limits of $f'(u)$ when $u > 1$ tends to 1 and when $u < 1$ tends to 1.

Indeed, the right derivative of f at 1 is equal to -1, while the left-derivative of f at 1 is equal to 1.

122. *Answer.* When $x \to 0$, $F(x)$ tends to $\frac{\pi}{2}f(0)$.

123. *Comments.* The intermediate value property of a derivative function f' (*i.e.*, the image by f' of any interval I is an interval of $\mathbb{R}$), be it continuous or not, is due to G. DARBOUX (1875).

So, one possible (and simplest) way to prove that a (non-continuous) function has no antiderivative is to check that it does not satisfy the intermediate value property.

124. *Comment.* Actually, neither f^2 nor g^2 is a derivative function... For g^2, this was proved directly by W. WILCOSZ in 1921. So, in contrast to continuous functions, for example, the product of derivative functions is not necessarily a derivative function.

For more on these questions, see

J.-B. Hiriart-Urruty, *Dérivation ou primitivation, quand tu nous tiens...* Bulletin de l'APMEP (Mathematics Teachers Association), n° 509, 340–348 (2014).

125. *Comment.* The result of question 2°) is due to Jessen (1929) (not Jensen...).

126. *Answers.* 1°) The answer is No. A counterexample is $f(x) = |x|$.

3°) The answer is Yes; in that case, the limit l is $f'(0)$.

127. *Comment.* The definition (1), proposed to approximate via an integral a notion of derivative of f at x, is due to C. Lanczos (1956).

128. *Answer.* Integrating over $[a,b]$ the inequality (2) yields the second inequality (on the right) in (1).

Next, decompose $\int_a^b f(x)dx$ as:

$$\int_a^b f(x)\ dx = \int_a^{\frac{a+b}{2}} f(x)\ dx + \int_{\frac{a+b}{2}}^b f(x)\ dx. \tag{3}$$

Change the variables $x = \frac{a+b}{2} - \frac{b-a}{2}t$ (resp. $x = \frac{a+b}{2} + \frac{b-a}{2}t$) in the first integral (resp. in the second integral) of the decomposition in (3), to obtain:

$$\frac{1}{b-a}\int_a^b f(x)\ dx = \frac{1}{2}\left\{\int_0^1 \left[f\left(\frac{a+b}{2} - \frac{b-a}{2}t\right) + f\left(\frac{a+b}{2} + \frac{b-a}{2}t\right)\right]\right\}\ dt.$$

Finally, the last step to obtain the first inequality (on the left) in (1) is to make use of the inequality $\frac{f(u)+f(v)}{2} \geqslant f\left(\frac{u+v}{2}\right)$.

Comments. - (1) is known as the Hermite–Hadamard inequality for convex functions.

- Each of the two sides of inequality (1) actually characterizes convex functions: If $f : I \to \mathbb{R}$ is a continuous function on the open interval I whose restriction to every interval $[a,b] \subset I$ satisfies the left inequality (resp. the right inequality) in (1) then f is convex on I.

129. *Answer.* The function f itself should be polynomial.

130. *Answer.* If $k \neq \frac{1}{2}$, only polynomial functions of degree 2 satisfy property (1).

If $k = \frac{1}{2}$, only polynomial functions of degree 3 satisfy property (1).

131. *Answer.* The only functions satisfying (1) are of the form $x \in \mathbb{R} \mapsto f(x) = ax^2 + b$, where a and b are real numbers.

133. *Comments.* f_h is sometimes called the STEKLOV function associated with f.

Indeed, when f is convex, $f \leqslant f_h$. Moreover, due to the continuity of f, the function f_h is continuously differentiable on $\mathbb{R}$. Therefore, $(f_{1/n})_{n \geqslant 1}$ is a sequence of C^1 convex functions converging pointwise to f, hence converging to f uniformly on every compact interval $[a, b] \subset \mathbb{R}$.

134. *Comments.* - Note that $(f^\Diamond)^\Diamond$ is f; hence, $(\cdot)^\Diamond$ is an involution operation.

- We have proved that f is convex if and only if $f^\Diamond$ is convex. Said otherwise: let $g : (0, +\infty) \to \mathbb{R}$; then $x \mapsto g(\frac{1}{x})$ is convex if and only if $x \mapsto xg(x)$ is convex.

- An interesting extension is to collect (or, even, to characterize) convex functions $f : (0, +\infty) \to \mathbb{R}$ for which $f^\Diamond = f$.

135. *Answer.* 1°) $f(x) = \sqrt{x}$; $f(x) = \ln(1 + x)$.

136. *Answer.* 2°) No, for the last question. For this, consider $f_n(x) = x^n$ on $[0, 1]$; the limit function f is indeed convex on $[0, 1]$, but the convergence of (f_n) towards f is not uniform $[0, 1]$.

137. *Comment.* The rather surprising property (1) does not extend beyond first-order differentiability.

138. *Answer.* The functions $f(x) = \cos(rx)$ and $f(x) = \cosh(rx)$, where r is any real number.

139. *Comment.* A result in the same vein is as follows: the only twice continuously differentiable (or even continuous) functions $f : \mathbb{R} \to \mathbb{R}$ satisfying

$$f(x + y)f(x - y) = [f(x)]^2 - [f(y)]^2 \text{ for all } x \text{ and } y \text{ in } \mathbb{R} \qquad (1')$$

are $x \mapsto Ax, A\sin(Bx), A\sinh(Bx)$ where A, B are real numbers. Such functions satisfy (1), and it is easy to illustrate (2) with them.

140. *Answer.* Recursion rules: $a_k^{(n+1)} = ka_k^{(n)} + a_{k-1}^{(n)}$; $a_{n+1}^{(n+1)} = a_n^{(n)}, a_1^{(n+1)} = a_1^{(n)}$. The first coefficients are:

$$\text{for } n = 2 : 1, 1;$$
$$\text{for } n = 3 : 1, 4, 1;$$
$$\text{for } n = 4 : 1, 17, 8, 1.$$

A general formula can be obtained for $a_k^{(n)}$:

$$a_k^{(n)} = \frac{1}{k!} \sum_{p=0}^{k} (-1)^p \binom{k}{p} (k - p)^n.$$

141. *Answer.* Recursion rules: $a_k^{(n+1)} = -na_k^{(n)} + a_{k-1}^{(n)}$; $a_{n+1}^{(n+1)} = a_n^{(n)}, a_1^{(n+1)} = -na_1^{(n)}$.

142. *Answer.* 2°) We have, for all nonnegative integers n:

$$g^{(n)}(x) = \int_0^1 t^n f^{(n+1)}(tx) \, dt \text{ for all } x \in \mathbb{R}.$$

In particular,

$$g^{(n)}(x) \to g^{(n)}(0) = \frac{f^{(n+1)}(0)}{n+1} \text{ as } x \to 0.$$

144. *Answer.* 2°) The answer is No. Take for example $f(t) = \cos t + 1$; then $F(t) = \sin t + t$ is not periodic.

145. *Answer.* The only point minimizing $I(r)$ on $\mathbb{R}$ is $r^* = f\left(\frac{a+b}{2}\right).$

146. *Answer.* There are no such functions. Two ways of proving this:

- Use the CAUCHY–SCHWARZ inequality to obtain a contradiction.

- Write

$$\mu^2 \int_0^1 f(x) \, dx - 2\mu \int_0^1 x f(x) \, dx + \int_0^1 x^2 f(x) \, dx = 0$$

$$= \int_0^1 f(x) \, (\mu - x)^2 dx.$$

147. *Answer.* 1°) Yes. 2°) $\nabla g(x,x) = \frac{1}{2} \begin{pmatrix} f''(x) \\ f''(x) \end{pmatrix}.$

148. *Comment.* It is amazing that one is able to calculate $\exp(tA)$ exactly in all cases. The only mathematical process required is how to calculate the eigenvalues of A; this therefore amounts to calculating the roots of a polynomial of degree 2.

149. *Comments.* - Beware that the powers for A and t in the two expressions in (1) are not the same.

- The result is clearly the "matricial cousin" of what is known for the scalar CAUCHY problem:

$$(c) \quad \begin{cases} y''(t) = a \, y(t), \\ y(0) = m_0, \\ y'(0) = m_1, \end{cases}$$

where $a \neq 0, m_1$ and m_2 are given real numbers. Beware of the sign of a, however. For example, if $a > 0$, the functions defined in (1) are $c(t) = \cosh(\sqrt{a}t)$ and $s(t) = \frac{\sin(\sqrt{a}t)}{\sqrt{a}}$.

151. *Answer.* 3°) In the ring $\mathcal{M}_2(\mathbb{R})$, the two matrices

$$\begin{bmatrix} 0 & 1 \\ 0 & 0 \end{bmatrix} \text{ and } \begin{bmatrix} 0 & 0 \\ 1 & 0 \end{bmatrix}$$

are nilpotent, but their sum, an invertible matrix, is not.

152. *Comments.* Another formulation of (1) is as follows: with $p = \frac{a+b+c}{2}$ (half of the perimeter of the triangle),

$$\mathcal{A} = \frac{1}{4}\sqrt{p(p-a)(p-b)(p-c)}. \tag{2}$$

This is known as HERON's formula, a beautiful formula indeed, allowing us to obtain $\mathcal{A}$ solely from $a, b, c...$, and valid for any triangle.

A further alternate formulation of (1) is

$$\mathcal{A}^2 = \frac{1}{16}\left[(a^2 + b^2 + c^2)^2 - 2(a^4 + b^4 + c^4)\right]. \tag{3}$$

153. *Answer.* The function f is maximized at the center of mass (or isobarycenter) of the triangle. The maximal value of f is

$$\max_{P \in \mathcal{T}} f(P) = \frac{8\mathcal{A}^3}{27AB \times BC \times AC}.$$

154. *Answer.* 1°) We have

$$\frac{1}{(AH)^2} = \frac{1}{(AB)^2} + \frac{1}{(AC)^2}.$$

2°) The line MN remains tangent to a circle centered at O and of radius $r = \frac{ab}{\sqrt{a^2+b^2}}$.

158. *Answer.* The dimension of $\mathcal{V}$ is $n^2 - 1$.

159. *Comment.* For each p, there are indeed subspaces V of dimension p for which (2) holds true. For this, V should be taken to be stable under the linear mapping

$$(x_1, x_2, \dots, x_n) \mapsto (x_2, x_3, \dots, x_n, x_1).$$

160. *Comment.* The set $\Omega \subset \mathcal{M}_n(\mathbb{C})$ of invertible matrices is an open dense subset in $\mathcal{M}_n(\mathbb{C})$. It is not a vector space, it is not connected.

161. *Answer.* Using Linear algebra. Proof by contradiction: suppose that all matrices in $\mathcal{H}$ are singular, and let us obtain a contradiction. $\mathcal{H}$ is the kernel of a non-null linear form $u : \mathcal{M}_n(\mathbb{R}) \to \mathbb{R}$. Since I_n is assumed not to belong to $\mathcal{H}$, we may assume, without loss of generality, that $u(I_n) = 1$.

Let $A \in \mathcal{M}_n(\mathbb{R})$ be a nilpotent matrix; then $A - u(A)I_n \in \ker u = \mathcal{H}$, hence is a singular matrix. In other words, $u(A)$ is an eigenvalue of A. But all the eigenvalues of nilpotent matrices are null (see Tapa 164, if necessary); that would mean that all the nilpotent matrices A are in $\mathcal{H}$. This is impossible: a sum of nilpotent matrices might not be in (the vector space) $\mathcal{H}$; for example, the following two matrices

$$A = [a_{i,j}] \text{ with } a_{n,1} = 1, \ a_{i,j} = 0 \text{ otherwise};$$
$$B = [b_{i,j}] \text{ with all } b_{i,j+1} = 1, \ b_{i,j} = 0 \text{ otherwise},$$

are nilpotent ($A^2 = 0$, $B^n = 0$) while their sum $A+B$ is invertible (since $\det(A+B) = (-1)^{n-1}$) and therefore could not be in $\mathcal{H}$.

163. *Answers.* Yes in the first case (characteristic polynomial), No in the second one (minimal polynomial). Here is an easy counterexample for the minimal polynomial: let

$$A(x) = \begin{bmatrix} x & 0 \\ 0 & 1 \end{bmatrix} \in \mathcal{M}_2(\mathbb{C}), \text{ with } x \in \mathbb{C}.$$

Look at what happens when $x \to 1$.

164. *Comment.* An extension of the results displayed above is as follows. Let M and N be in $\mathcal{M}_n(\mathbb{C})$; we then have:

$$\left(\operatorname{tr}(M^k) = \operatorname{tr}(N^k) \text{ for all positive integers } k \right)$$
$$\Leftrightarrow \tag{2}$$
$$(M \text{ and } N \text{ have the same characteristic polynomial}).$$

165. *Answer.* We have:

$$\Delta_n = (-1)^n \times \frac{f^{n+1}}{n!} \times \left(\frac{1}{f}\right)^{(n)}. \tag{3}$$

167. *Answers.* The set X_n is always closed.

It is bounded only when $n = 1$. Indeed:

- the set X_2 contains unbounded matrices like $\begin{bmatrix} 0 & r \\ \frac{1}{r} & 0 \end{bmatrix}, r > 0$;

- the set $X_n, n \geqslant 3$, contains unbounded matrices of the form $\begin{bmatrix} 0 & r & \\ \frac{1}{r} & 0 & \\ & & I_{n-2} \end{bmatrix}, r > 0.$

168. *Answers.* 1°) There are $n!$ of them.

2°) The solution matrices A are of the form $aI_n + bE$, where E is the matrix all of whose entries are equal to 1, and a, b are real numbers. In short, A has c as diagonal entries and b as non-diagonal entries.

Comment. The matrices which commute with all the permutation matrices are also those which commute with all the bistochastic matrices. A matrix $M = [m_{ij}] \in \mathcal{M}_n(\mathbb{R})$ is called bistochastic when its entries m_{ij} satisfy:

$$m_{ij} \geqslant 0 \text{ for all } i, j$$

$$\sum_{i=1}^{n} m_{ij} = \sum_{j=1}^{n} m_{ij} = 1 \text{ for all } i, j.$$

It turns out that any bistochastic matrix is a convex combination of permutation matrices (a theorem of G. BIRKHOFF).

169. *Answer.* 1°) (b) Equality (1) is not true for $D = 0$ and n odd. The simplest counterexample is therefore with $\begin{bmatrix} a & b \\ c & 0 \end{bmatrix} \in \mathcal{M}_2(\mathbb{R})$.

Comment. A more general formula than (1), whether D is invertible or not, is:

$$\det M = (-1)^{n-rank(D)} \det(AD^T + BC^T). \tag{3}$$

173. *Answer.* In the fourth question, $\exp(C)$ is I_2, but the only symmetric matrix C for which $\exp(C)$ is I_2 is the null matrix.

174. *Comment.* According to the previous proposal (Tapa 173), $\exp(A)$ is a rotation in $\mathbb{R}^3$. Since the trace of $\exp(A)$ is $1 + \cos r$, the cosine of the angle of the rotation is $\cos r$.

175. *Answer.* The answer seems obvious... but it has to be proved. The limit function f is indeed a quadratic form on $\mathbb{R}^n$. Actually, for all x in $\mathbb{R}^n$, $f(x) = x^T A_\infty x$, where $A_\infty \in \mathcal{M}_n(\mathbb{R})$ is symmetric, and the sequence of matrices (A_k) converges to A_∞ in $\mathcal{M}_n(\mathbb{R})$.

176. *Answers.* The inertia of q_1 is $(1, n-1, 0)$, that is to say: Q_1 has 1 positive eigenvalue, $n - 1$ negative eigenvalues, and 0 null eigenvalues.

The inertia of q_2 is $(n, 0, 0)$: q_2 is therefore positive definite on $\mathbb{R}^n$.

178. *Comments.* (a) (1) is called the DUNKL–WILLIAMS inequality.

(b) We deduce from (1) that the mapping $\nu : x \mapsto \nu(x) = \frac{x}{\|x\|}$ is 1-LIPSCHITZ on the set $\Omega = \{x : \|x\| \geqslant 1\}$. This does not hold true for any norm on $\mathbb{R}^n$.

(c) (2) is an improvement of the usual CAUCHY–SCHWARZ inequality since

$$-1 \leqslant 1 - 2\left(\frac{\|x - y\|}{\|x\| + \|y\|}\right)^2 \text{ and } -1 + 2\left(\frac{\|x + y\|}{\|x\| + \|y\|}\right)^2 \leqslant 1.$$

179. *Comments.* (*a*) There is no loss of generality in assuming that both S and T are closed and symmetric ($S = -S; T = -T$).

(*b*) The result of the proposal may fail if more than two matrices are involved, even with $n = 2$.

182. *Comment.* When $n = 1$, one recognizes above a result learnt in high school: the trinomial $ax^2 + 2bx + c$ (with $a \neq 0$) is nonnegative on $\mathbb{R}$ if and only if ($a > 0$ and $\Delta = b^2 - ac \leqslant 0$). This is precisely the test for the matrix $\begin{bmatrix} a & b \\ b & c \end{bmatrix}$ to be positive semidefinite. Here, $\Delta = b^2 - ac$ is the so-called (reduced) *discriminant* of the trinomial, and is also the negative of the *determinant* of the matrix $\begin{bmatrix} a & b \\ b & c \end{bmatrix}$.

183. *Answer.* The signature of q is $\left(\frac{n(n+1)}{2}, \frac{n(n-1)}{2}, 0 \right)$.

184. *Comment.* Since AB is not necessarily symmetric ($(AB)^T \neq BA$), the "symmetrized product" of A and B, namely $S = AB + BA$, is called. Then note that $x^T S x = 2x^T(AB)x = 2x^T(BA)x$.

185. *Answer.* A counterexample for the second question is:

$$A = \begin{bmatrix} 1 & -2 \\ -2 & 1 \end{bmatrix} \in \mathcal{M}_2(\mathbb{R})$$

and $H = \mathbb{R} \times \{0\}$.

Comments. - The result extends to a closed convex cone K of $\mathbb{R}^n$ (instead of a vector space H); then one considers, instead of $H^\perp$, the so-called polar cone $K^\ominus$ of K defined as follows:

$$K^\ominus = \left\{ y \in \mathbb{R}^n \mid y^T x \leqslant 0 \text{ for all } x \text{ in } K \right\}.$$

This resulting characterization, with closed convex cones (of a higher difficulty, ★★★), is due to HAN and MANGASARIAN (1984).

- Observe the beautiful symmetry in the formulation of the two conditions in (1) since $(A^{-1})^{-1} = A$ and $(H^\perp)^\perp = H$ (and also $(K^\ominus)^\ominus = K$ in the generalization).

186. *Answers.* 1°) Yes, C is positive semidefinite.

2°) C is positive definite whenever A and B are positive definite.

187. *Answers.* 1°) $\bar{u} = (A + B)^{-1} Bx$ and $\bar{v} = (A + B)^{-1} Ax$.

2°) $\mathcal{A} = (A^{-1} + B^{-1})^{-1} = A(A + B)^{-1} B = B(A + B)^{-1} A$.

Comments. - The relation $A // B = (A^{-1} + B^{-1})^{-1}$ has an interesting physical interpretation, by considering an electrical circuit made up of two (so-called) generalized resistors (represented by the matrices) A and B, and connected in parallel. Then, $\mathcal{A} = A//B$ plays the role of a generalized resistor equivalent to A and B connected in parallel; when $n = 1$, we get the more familiar relation on resistances $r_1 > 0$ and $r_2 > 0$ put in parallel: $\frac{1}{r_1//r_2} = \frac{1}{r_1} + \frac{1}{r_2}$.

- A long problem on the parallel addition of matrices was posed by the author in a national competition for entering one of the prestigious French engineering schools (Ecole Polytechnique) in 1999. Its presentation and a detailed solution can be found in pages 132–144 of the book

D. Azé and J.-B. Hiriart-Urruty, *Analyse variationnelle et optimisation. Eléments de cours, exercices et problèmes corrigés*. Cepaduès Editions (2010).

- For a recent pedagogical survey on symmetric positive semidefinite and positive definite matrices, see

J.-B. Hiriart-Urruty and J. Malick, *A fresh variational-analysis look at the positive semidefinite matrices world*. J. of Optimization Theory and Applications 153 (3), 551–577 (2012).

188. *Answer.* 2°) The infimum in (1) is $x^T \left[\mathbb{E}(A)\right]^{-1} x$. The unique $X^* \in L^1(x)$ where this infimum is achieved is: $X^*(\omega) = A(\omega) \left[\mathbb{E}(A)\right]^{-1} x$.

195. *Comment.* A more subtle result of Motzkin and Taussky (1955) asserts the following (of a higher difficulty, ★★★): let A and B be two diagonalizable matrices in $\mathcal{M}_n(\mathbb{C})$; then the following equivalence holds:

$$(AB = BA) \iff (zA + z'B \text{ is diagonalizable for all } z, z' \text{ in } \mathbb{C}).$$

197. *Answer.* Let λ_1 (resp. λ_n) denote the largest (resp. the smallest) eigenvalue of A. The maximal value of $f(x)$ over the unit sphere of $\mathbb{R}^n$ is $\left(\frac{\lambda_1 - \lambda_n}{2}\right)^2$.

198. *Answers.* Let λ_1 (resp. λ_n) denote the largest (resp. the smallest) eigenvalue of A. The maximal value of $f(x)$ over the unit sphere of $\mathbb{R}^n$ is $\left(\sqrt{\lambda_1} - \sqrt{\lambda_n}\right)^2$. This maximal value is attained at

$$x = \left(\frac{\sqrt{\lambda_1}}{\sqrt{\lambda_1} + \sqrt{\lambda_n}}\right)^{1/2} u \pm \left(\frac{\sqrt{\lambda_n}}{\sqrt{\lambda_1} + \sqrt{\lambda_n}}\right)^{1/2} v$$

where u and v are unit eigenvectors corresponding to λ_1 and λ_n, respectively.

201. *Answers.* The matrix H, introduced by the German geophysicist F.R. Helmert is indeed orthogonal; it is used in Statistics.

The eigenvalues of $A_{a,b}$ are: $\lambda_1 = a + (n-1)b; \lambda_2 = \lambda_3 = ... = \lambda_n = a - b$. The matrix H helps to diagonalize $A_{a,b}$, independently of a and b.

202. *Answer.* A detailed proof can be found in:

> J.-B. HIRIART-URRUTY, *Le théorème déterminantal en bref.* Revue de la Filière Mathématiques RMS (ex-Revue de Mathématiques Spéciales), n° 119, 15–22 (2008–2009).

204. *Comment.* According to the result of the second question, $\mathcal{O}_2^+(\mathbb{R})$ and $\mathcal{O}_2^-(\mathbb{R})$ geometrically look like two circles placed in two orthogonal planes.

205. *Comments.* - Any $M \in \mathcal{SO}_n(\mathbb{R})$ not admitting -1 as an eigenvalue can be factorized, in a unique way, as $M = (I_n - A)(I_n + A)^{-1}$ with A antisymmetric. This technique is called the CAYLEY parametrization of matrices in $\mathcal{SO}_n(\mathbb{R})$.

- After some "homogenization" of parameters, passing from any vector $(a, b, c) \in \mathbb{R}^3$ to unitary vectors $(\alpha, \beta, \gamma, \delta) \in \mathbb{R}^4$, one can rewrite the expression (1) in a better-looking form. Indeed, the $M \in \mathcal{SO}_3(\mathbb{R})$ are of the form:

$$M = \begin{bmatrix} \alpha^2 - \beta^2 - \gamma^2 + \delta^2 & 2(\alpha\beta - \gamma\delta) & 2(\alpha\gamma + \beta\delta) \\ 2(\alpha\beta + \gamma\delta) & -\alpha^2 + \beta^2 - \gamma^2 + \delta^2 & 2(\beta\gamma - \alpha\delta) \\ 2(\alpha\gamma - \beta\delta) & 2(\beta\gamma + \alpha\delta) & -\alpha^2 - \beta^2 + \gamma^2 + \delta^2 \end{bmatrix}, \quad (2)$$

where $(\alpha, \beta, \gamma, \delta) \in \mathbb{R}^4$ satisfies $\alpha^2 + \beta^2 + \gamma^2 + \delta^2 = 1$.

206. *Comment.* A possible generalization is as follows: let $F : x \in \mathbb{R}^n \mapsto F(x) \in \mathbb{R}^n$ be a twice continuously differentiable mapping; with the assumption that the Jacobian matrix $JF(x)$ is orthogonal for all $x \in \mathbb{R}^n$, we would get that F is an affine isometry of $\mathbb{R}^n$.

207. *Answer.* We expand $\|A - B\|^2$ and get successively:

$$\begin{aligned} \|A - B\|^2 &= 2n - \text{tr}(A^T B + B^T A); \\ pq \, \text{tr}(A^T B + B^T A) &= n(1 - p^2 - q^2); \\ \|A - B\| &= 0. \end{aligned}$$

Thus $A = B$ necessarily. In other words, there cannot be any orthogonal matrix in the line-segment joining A to B, except at the end-points A and B.

208. *Answer.* 3°) The unit ball for the norm $\|\cdot\|_{(k)}$ is a convex compact polyhedral set; S is the part of the unit sphere for $\|\cdot\|_{(2)}$ lying in the nonnegative orthant of $\mathbb{R}^3$. The relevant points to draw S are the four vertices $(0, 0, 1)$, $(0, 1, 0)$, $(0, 0, 1)$ and $(\frac{1}{2}, \frac{1}{2}, \frac{1}{2})$.

209. *Comment.* As expected, inequality (2), obtained under a stronger assumption on u and v, is sharper than inequality (1).

212. *Comment.* The inequality (1) generalizes to $\mathbb{R}^n$, with $\frac{r}{n}$ as a corresponding constant.

213. *Answers.* 1°) For a polynomial function P of the real variable, there are three possibilities for its range $P(\mathbb{R})$: just a point, the whole of $\mathbb{R}$, or a closed unbounded interval $[m, +\infty)$ or $(-\infty, m]$.

2°) Let $P(x, y) = (xy - 1)^2 + x^2$. Then the range $P(\mathbb{R}^2)$ of $\mathbb{R}^2$ by P is the open interval $(0, +\infty)$.

215. *Comments.* Question 2°): The roots of P' are exactly the two focuses of the ellipse of largest area contained in the triangle $\mathcal{T}$ (called the STEINER ellipse); this result is known as MARDEN's theorem (1965) or the SIEBECK–MARDEN theorem (1864).

Question 3°): This result is known as the GAUSS–LUCAS theorem.

216. *Comments.* - An illustration of the result presented above is as follows. Suppose that the temperature at the location P on the terrestrial globe is a continuous function of the coordinates of P. There then are two opposite (or antipodal) points on a meridian line of the globe which are at the same temperature.

- The intermediate value theorem or fixed point theorem for continuous mappings come into play to solve some geometrical problems like: the existence of a straight line cut that divides a pizza exactly in half; the existence of a square box that captures perfectly a compact planar region.

217. *Comments.* - In the right-hand side of (1), one cannot substitute "critical point of f" (*i.e.*, a point x^* at which $f'(x^*) = 0$) for "local minimizer of f"; consider as a counterexample $f(x) = x^3$ and $\overline{x} = 0$.

- A global minimizer $\overline{x}$ is indeed a local minimizer, but (1) pinpoints the difference: if $\overline{x}$ is a local minimizer which is not a global minimizer, there exists another point x^*, at the same level as $\overline{x}$ (*i.e.*, $f(x^*) = f(\overline{x})$), which is not a local minimizer.

- The continuity assumption in the result above cannot be removed: for the function $x \in \mathbb{R} \mapsto f(x) = 1$ if $x \neq 0$, $f(0) = 0$, every point x^* is a local minimizer but none of the $x^* \neq 0$ is a global minimizer.

218. *Comment.* The counterexample proposed here was brought up by G. PEANO (1884).

219. *Answers.* 1°) Yes; the gradient of f is necessarily null everywhere, so the function is constant.

2°) The answer is still Yes. We can even relax the assumption by just assuming that all points $x \in \mathbb{R}^n$ are either local minimizers or local maximizers of f (a bit more difficult to prove, ★★★).

3°) The answer is No. Counterexample: $f : \mathbb{R} \to \mathbb{R}$ with $f(0) = 0, f(x) = 1$ for all $x \neq 0$.

4°) The function f is constant, without assuming any property on it.

220. *Answer.* A detailed solution can be found in

J.-B. HIRIART-URRUTY, *A short proof of the variational principle for approximate solutions of a minimization problem.* Amer. Math. Monthly 90 (1983), 206–207.

Comment. The necessary conditions for approximate optimality, such as expressed in (1)-(2)-(3), are due to I. EKELAND (1973), in a much more general context.

223. *Answer.* There is only one possibility: $C = \mathbb{R}^n$.

224. *Answer.* This maximal dimension of a vector space contained in Φ is $\max(m, n)$.

225. *Answer.* 1°) Yes. Note, however, that neither the connectedness nor the closedness assumption on G_f can be (completely) removed.

226. *Answer.* 2°) Let $f(x_1, x_2, ..., x_n) = (x_1^2, 0, ..., 0)$. This mapping f is neither affine nor injective, but it transforms any convex set into a convex set.

228. *Answer.* The only critical point of f is $(a, b) = (1, 0)$. It is a strict local minimizer of f since the Hessian matrix (or matrix of second partial derivatives) of f at $(1, 0)$ is $6 \begin{bmatrix} 2 & -1 \\ -1 & 1 \end{bmatrix}$, which is positive definite.

But, $f(x, 0) = 2x^3 - 6x + 3$ is neither bounded from below nor bounded from above.

Comments. The situation described above could not happen with functions f of a single variable.

It could happen with polynomial functions $f : \mathbb{R}^2 \to \mathbb{R}$, like

$$f(x, y) = x^2(1 + y)^3 + y^4,$$
or
$$f(x, y) = x^2(1 + y)^3 + y^2.$$

The only critical point here is $(a, b) = (0, 0)$. Substituting y^2 for y^4 furthermore means that the Hessian matrix (or of second partial derivatives) of f at $(0, 0)$ is positive definite.

No example of such a "strange minimization problem" could be exhibited with polynomial functions of degree $\leqslant 4$.

229. *Answers.* 1°) The infimum of f on $\mathbb{R}^2$ is 0 and it is not attained. To see why, consider $f(\frac{1}{x}, x)$ with $x \to +\infty$.

2°) The only critical point of f is $(0, 0)$ and, since the Hessian matrix (or of second partial derivatives) of f at $(0, 0)$ is $2 \begin{bmatrix} 1 & -1 \\ -1 & 0 \end{bmatrix}$ (an indefinite symmetric matrix), $(0, 0)$ is a saddle point.

140

To check that $(0,0)$ is neither a local maximizer nor a local minimizer, one could also look at $f(x,x)$ and $f(x,-x)$ for x in the interval $(-1,1)$ and compare it to $f(0,0) = 1$.

Comment. Another example of this type is

$$f(x,y) = (x-y)^2 + \left(x^2 - y^2 - 1\right)^2.$$

Consider here $f(\cosh x, \sinh x)$ for $x \to +\infty$.

As a general rule, one can consider $f(x,y) = [P(x,y)]^2 + [Q(x,y)]^2$, where P and Q are two polynomial functions such that the curves of equations $P(x,y) = 0$ and $Q(x,y) = 0$ are disjoint but of zero mutual distance.

230. *Answer.* The two global minimizers of f are $(1,2)$ and $(-1,0)$.

231. *Answer.* This locus is formed by two symmetric parabolas, with equations $y = \pm\frac{x^2-1}{2}$.

233. *Answer.* The surface area of Σ is $4a^2 \times \arcsin\left(\frac{b}{a}\right)$. When $b = a$, the parachute is a half-sphere, whose surface area is $2\pi a^2$.

234. *Answer.* 2°) We have: $vol(S) = 9\pi(2 - \sqrt{2})$ unit$^3 \simeq 113.098$ unit^3.

 3°) The proportion is about 14.65%.

235. *Answer.* The volume of S is $\frac{8}{9}(3\pi - 4)r^3 \simeq 4.82r^3$.

 Comment. The curve drawn on the half-sphere $(x^2 + y^2 + z^2 = 4r^2, z \geqslant 0)$ by the cylinder $x^2 + (y-r)^2 = r^2$ is called the window of V. VIVIANI (1622–1703).

236. *Answers.* 2°) The surface area of Σ_0 is surprisingly simple, it is just r^2. Hence, the surface area of the whole Σ is $8r^2$.

 3°) The volume of S_0 is $\frac{2}{3}r^3$. Hence, the volume of the whole hard candy S is $\frac{16}{3}r^3 \simeq 5.33r^3$.

 Comments. - The solid defined in (1) sometimes is called the STEINMETZ solid, even though it was known long before C.P. STEINMETZ studied it.

 - As expected, one may consider the solid obtained by intersection of $3, 4, \ldots, 100$ cylinders. The first step is then to create a good drawing of the resulting solid (using a computer). For example (case of 3 cylinders), the volume of the solid defined as

$$\left\{(x,y,z) \ : \ x^2 + y^2 \leqslant r^2, \ x^2 + z^2 \leqslant r^2, \ y^2 + z^2 \leqslant r^2\right\}$$

can be calculated by using cylindrical coordinates; it is $8(2 - \sqrt{2})r^3 \simeq 4.68r^3$.

As the number of cylinders increases, the solid obtained by intersection is increasingly spherical; for example, already with $n = 100$ it is hard to see the difference with a sphere.

141

237. *Answers.* 2°) - Coordinates of P : $x = \frac{1}{3}(1 - 4\sqrt{2}) \simeq -1.5522$, $y = \frac{2}{3}(1 + 2\sqrt{2}) \simeq$ 2.5522. Coordinates of Q : $x = \frac{1}{3}(1 + 4\sqrt{2}) \simeq 2.2189$, $y = -\frac{2}{3}(2\sqrt{2} - 1) \simeq -1.2189$.

- Equation of Δ_1 : $Y = X + \frac{1}{3} + \frac{8\sqrt{2}}{3}$; equation of Δ_2 : $Y = X + \frac{1}{3} - \frac{8\sqrt{2}}{3}$. Hence they are parallel.

- The distance between the parallel lines Δ_1 and Δ_2 is $d = \frac{16}{3}$.

3°) The length of the curve Γ is $\frac{16}{3}\pi$, the same as a circle of diameter d.

Comments. - The oval whose boundary is the smooth curve Γ turns out to be "of constant width"; that means that two parallel tangent lines to the oval are always at the same mutual distance d (like for a circle of diameter d). Indeed, the two points corresponding to parameters shifted by 2π, P corresponding to t, Q corresponding to $t + 2\pi$, are always at the same mutual distance $\frac{16}{3}$, and the tangent lines to Γ at P and Q are parallel.

- For more on these ovals or solids "of constant width", see

T. BAYEN and J.-B. HIRIART-URRUTY, *Objets convexes de largeur constante (en 2D) ou d'épaisseur constante (en 3D) : du neuf avec du vieux*. Annales des sciences mathématiques du Québec. Vol. 36, n° 1, 17–42 (2012).

238. *Answers.* 1°) (b). The depth is maximal for $(x, y) = (0, 0)$; this maximal depth is h_M.

(c) The area of the lake is πab.

2°) After a first change of variables $(u = \frac{x}{a}, y = \frac{y}{b})$, we then pass to polar coordinates. The integral to be calculated is

$$V = h_M \times ab \times \int_0^{2\pi} \left[\int_0^1 \cos(\frac{\pi r}{2}) r \, dr \right] d\theta.$$

Finally,

$$V = h_M \times ab \times 4(1 - \frac{2}{\pi}) \simeq h_M \times ab \times 1.4535.$$

239. *Answers.* 1°) (a) Σ is a piece of a hyperbolic paraboloid (of equation $z = -x^2 + y^2 + 2$).

(b) We obtain:

$$4 \leqslant \text{vol}(S) \leqslant 12.$$

(c) The intersection of Σ with the plane of equation $z = 2$ is constituted of two line segments (the two rafters):

$$\left(\begin{matrix} x = y, \ z = 2 \\ -1 \leq x \leq 1, \ -1 \leq y \leq 1 \end{matrix} \right) \text{ and } \left(\begin{matrix} -x = y, \ z = 2 \\ -1 \leq x \leq 1, \ -1 \leq y \leq 1 \end{matrix} \right).$$

2°) (a) $D = \{(x, y) \in \mathbb{R}^2 \mid -1 \leq x \leq 1 \text{ and } -1 \leq y \leq 1\}$; $f(x, y) = -x^2 + y^2 + 2$.

(b) $V = 8$.

3°) (b) $\mathcal{A} = \frac{\pi(5\sqrt{5}-1)}{6} \simeq 5.330$.

Comment. There are indeed sports centres shaped like S; spectators can see the game (running on the ground), but not necessarily the other spectators who are in the opposite terraces.

240. *Answers.* 1°) (b). The height h is 4.

2°) (a). $D = \{(x, y) \in \mathbb{R}^2 \mid x^2 + y^2 \leq 9\}$; $f(x, y) = \sqrt{25 - x^2 - y^2}$.

(b) $V = \frac{122\pi}{3} \simeq 127.758$.

3°) (b). $\mathcal{A} = 10\pi \simeq 31.416$; it represents about 20% of the area of the half-sphere.

241. *Answers.* 2°) $l = 8r$. 3°) $\mathcal{A} = 3\pi r^2$.

4°) (a) For symmetry reasons, $x_G = \pi r$.

(b) $y_G = \frac{4r}{3}$.

242. *Answers.* 4°) The Cartesian equation of Γ_4 in the new basis $(O; \vec{u}, \vec{v})$ is

$$X \in \left[-\frac{\sqrt{2}}{2}, -\frac{\sqrt{2}}{2} \right] \mapsto Y = \frac{\sqrt{2}}{4}(1 + 2X^2).$$

Accordingly, $l(\Gamma_4) = 1 + \frac{\sqrt{2}}{2} \ln(1 + \sqrt{2}) \simeq 1.88$.

5°) $l(\Gamma_3) = \frac{3}{2} = 1.50$. The successive lengths of arcs could be compared:

$$l(\Gamma_1) \simeq 0.78, \quad l(\Gamma_2) \simeq 0.71, \quad l(\Gamma_3) = 1.50, \quad l(\Gamma_4) \simeq 1.88.$$

We have: $\mathcal{A}_3 = \frac{3\pi}{32} \simeq 0.294$.

Comment. By completing Γ_3 with the help of symmetries with respect to the Ox and Oy axes, we get a closed curve included in the box $[-1, 1] \times [-1, 1]$ called an *astroid*.

243. *Answers.* 2°) $V = \frac{\pi r^4}{4} \times \frac{z_2 - z_1}{z_1 z_2}$.

3°) $x_G = \frac{r^2}{4} \times \frac{z_1 + z_2}{z_1 z_2}$; $z_G = \frac{z_1 z_2}{z_2 - z_1} \times \log(\frac{z_2}{z_1})$, a coordinate independent of r! When z_2 tends to z_1, the point G converges towards $(\frac{r^2}{2z_1}, 0, z_1)$, which is the center of the limit-disk obtained by shrinking S.

244. *Answer.* 2°) The area of S is $\frac{49}{3}$.

3°) $x_G = y_G = \frac{837}{196} \simeq 4.270$.

245. *Answers.* 1°) Take $x(\cdot)$ as the constant function $x(a)$ and $y(t) = x(t)$ in (1), integrate between a and b, and use the formula $\int_a^b \frac{d\varphi}{dt} dt = \varphi(b) - \varphi(a)$.

2°) - Usual change of variables: make $f(t, x) = f(x)$ in (4).

- Integration by parts formula: make $f(t, x) = f(t)$ in (4), use the formula $\int_a^b c\, dt = c \times b - c \times a$, and rearrange the terms.

143

248. *Answers.* 1°) The gradient of f at P is the unit vector pointing towards the vertex A.

2°) The gradient of g at P is the unit vector pointing towards the side BC.

3°) The orthocenter is the only point in $\mathcal{T}$ where the gradient of the convex function h is the zero vector.

Comment. For more on this question of characterizing the centers of a triangle by optimization, see

J.-B. HIRIART-URRUTY and P.-J. LAURENT, *A characterization by optimization of the orthocenter of a triangle.* Elemente der Mathematik 70, 45–48 (2015).

249. *Answers.* 2°) (a) :

$$\frac{1}{\left\|\overrightarrow{OA_1}\right\|^2} + \frac{1}{\left\|\overrightarrow{OA_2}\right\|^2} + \frac{1}{\left\|\overrightarrow{OA_3}\right\|^2} = \frac{1}{a^2} + \frac{1}{b^2} + \frac{1}{c^2}.$$

2°) (b). The plane P remains tangent to a sphere centered at 0 and of radius $r = 1/\sqrt{\frac{1}{a^2} + \frac{1}{b^2} + \frac{1}{c^2}}$.

250. *Comments.* This result is attributed to J.-P. GUA DE MALVES (published in his memoirs of 1783). For more on this question, read the following two references:

J.-B. HIRIART-URRUTY, J.-P. QUADRAT and J.B. LASSERRE, *Pythagoras' theorem for areas.* American Math. Monthly, 549–551 (2001).

J.-M. LÉVY-LEBLOND, *The Pythagorean theorem extended and deflated.* The Mathematical Intelligencer, vol.7, n°2, 5–6 (2005).

251. *Answers.* 2°) We have:

$$a_0 = \frac{1}{\pi}; \ a_{2p} = \frac{2}{\pi(1 - 4p^2)} \text{ for } p \geqslant 1; \ a_{2p+1} = 0 \text{ for } p \geqslant 0.$$

$$b_1 = \frac{1}{2}; \ b_n = 0 \text{ if } n \geqslant 2.$$

4°)– 5°) We get:

$$\sum_{p=1}^{+\infty} \frac{1}{4p^2 - 1} = \frac{1}{2}; \ \sum_{p=1}^{+\infty} \frac{1}{(4p^2 - 1)^2} = \frac{\pi^2}{16} - \frac{1}{2} \simeq 0.1168.$$

Comments. - An interesting variant consists in considering the even function $g(x) = |\sin(x)|$. We then get

$$\sin(x) = \frac{2}{\pi} - \frac{4}{\pi} \sum_{p=1}^{+\infty} \frac{\cos(2px)}{4p^2 - 1} \text{ for all } x \in [0, \pi]. \tag{1}$$

This always surprises beginners and students since it expresses the sine function (on $[0, \pi]$) as a sum of... cosine functions.

- In all questions concerning FOURIER series (proposals 251, 252, 253, 254, 255, 331), the considered functions are piecewise smooth (*i.e.*, both f and f' are piecewise continuous), so that only "pocket theorems" like DIRICHLET's theorem will be called.

252. *Answers.* 2°)– 3°) We have:

$$f(x) = \frac{\pi^2}{6} - \sum_{p=1}^{+\infty} \frac{\cos(2px)}{p^2} \text{ for all } x \in \mathbb{R}.$$

4°)– 5°) We get:

$$\sum_{p=1}^{+\infty} \frac{(-1)^{P+1}}{p^2} = \frac{\pi^2}{12}; \sum_{p=1}^{+\infty} \frac{1}{p^4} = \frac{\pi^4}{90}.$$

253. *Answers.* 3°) We have, for all $x \in \mathbb{R}$:

$$\frac{f(x+) + f(x-)}{2} = \sum_{p=0}^{+\infty} \frac{4(-1)^p}{\pi(2p+1)} \cos\left[(2p+1)x\right].$$

4°)– 5°) We get:

$$\sum_{p=0}^{+\infty} \frac{(-1)^p}{2p+1} = \frac{\pi}{4}; \sum_{p=0}^{+\infty} \frac{1}{(2p+1)^2} = \frac{\pi^2}{8}.$$

254. *Answers.* 4°)– 5°) We get:

$$\sum_{p=0}^{+\infty} \frac{1}{(2p+1)^2} = \frac{\pi^2}{8}; \sum_{p=0}^{+\infty} \frac{1}{(2p+1)^2}(-1)^{\lfloor(p+1)/2\rfloor} = \frac{\pi^2\sqrt{2}}{16};$$

$$\sum_{p=0}^{+\infty} \frac{1}{(2p+1)^4} = \frac{\pi^4}{96}.$$

256. *Answers.* 1°) (*b*). It is necessary to distinguish two cases: $x = 0$ and $x \neq 0$. We have:

$$\begin{aligned} S(x) &= 1 + x^2 \text{ if } x \neq 0, \\ S(0) &= 0. \end{aligned}$$

2°) One reason is: the functions u_n are continuous on $\mathbb{R}$, the sum S is not continuous on $\mathbb{R}$; hence the convergence of the series of functions with general term $(u_n)_{n \geqslant 0}$ towards the function S cannot be uniform on $\mathbb{R}$.

Another reason is that one can calculate $S(x) - S_N(x)$ explicitly for all $x \in \mathbb{R}$; indeed

$$\sup_{x \in \mathbb{R}} |S(x) - S_N(x)| = 1 \nrightarrow 0 \text{ as } N \to +\infty.$$

Comment. It is interesting to draw graphs of some S_N to observe the "detachment" of them at the point $x = 0$.

257. *Comment.* The expression of $(r, \theta) = F^{-1}(x, y)$ depends on the open set of the plane where the points (x, y) are considered. The formula (2), a bit overlooked (if ones compares it with the more familiar (1-1)), has the advantage of being valid (and concise) for "almost all" the points in the plane.

258. *Answers.* 1°) The new partial differential equation is

$$\frac{\partial \widehat{u}}{\partial \rho} = 1. \tag{2}$$

A general solution of (2) is $\widehat{u}(\rho, \theta) = \rho + h(\theta)$, where h is a continuously differentiable function of the real variable θ.

2°) For solutions of (1), we finally obtain

$$u(x, y) = \sqrt{x^2 + y^2} + g\left(\frac{y}{x}\right),$$

where $g : (0, +\infty) \to \mathbb{R}$ is an arbitrary continuously differentiable function.

259. *Answer.* 2°) $u(x, t) = \frac{1}{c} \sin(ct) \times \sin(x)$.

Comment. If one is just interested in solutions of the partial differential equation $\frac{\partial^2 u}{\partial t^2} = c^2 \frac{\partial^2 u}{\partial x^2}$, a family of solutions is provided by

$$u(x, t) = f_1(x + ct) + f_2(x - ct),$$

where f_1 and f_2 are twice continuously differentiable functions of a real variable.

261. *Answer.* 2°) A family of solutions of (1) is given by

$$z(x, t) = e^{-t} f_1(x + t) + e^{-t} f_2(x - ct),$$

where f_1 and f_2 are twice continuously differentiable functions of a real variable.

262. *Answers.* 1°) We have
$$\mathcal{A}(a, b) = \pi a b. \tag{3}$$

146

2°) (a) Denoting by $E = \sqrt{1 - \left(\frac{b}{a}\right)^2}$ the so-called eccentricity of the ellipse $\mathcal{E}$, we have:

$$L(a,b) \;=\; \int_0^{2\pi} \sqrt{a^2 \sin^2(t) + b^2 \cos^2(t)} \; dt = 4 \int_0^{\frac{\pi}{2}} \sqrt{1 - E^2 \sin^2 \theta} \; d\theta; \tag{4}$$

$$L(a,b) \;=\; 2\pi a \left[1 - \left(\frac{1}{2}\right)^2 E^2 - \left(\frac{1}{2} \times \frac{3}{4}\right)^2 \frac{E^4}{3} - \left(\frac{1}{2} \times \frac{3}{4} \times \frac{5}{6}\right)^2 \frac{E^6}{5} + ... \right] \tag{5}$$

(a series of MacLaurin).

(b) We have:

$$2\pi \sqrt{ab} \leqslant \pi(a+b) \leqslant L(a,b) \leqslant \pi \sqrt{2(a^2 + b^2)}. \tag{6}$$

Comment. There are several other approximations of $L(a,b)$; one of them, due to RAMANUJAN, turns out to be very precise, at least when the eccentricity E is moderate. Here it is:

$$L(a,b) \approx \pi \left[(a+b) - \frac{3(a-b)^2}{10(a+b) + \sqrt{a^2 + 14ab + b^2}} \right]. \tag{7}$$

For more on the question of the length of an ellipse (history, various proposals, etc.), see

J.-B. HIRIART-URRUTY, *Propos elliptiques...*. Tangente, n°163, 44–46 (2015).

263. *Answer.* The distance $O\Omega$ constantly equals $\sqrt{a^2 + b^2}$; the locus Γ is a portion of the arc of the circle centered at O and of radius $\sqrt{a^2 + b^2}$. To be more precise,

$$\Gamma = \left\{ (x,y) : b \leqslant x \leqslant a; \; b \leqslant y \leqslant a; \; x^2 + y^2 = a^2 + b^2 \right\}.$$

Comment. The locus of other points in $\mathcal{E}$ could be complicated curves in the positive quadrant; try some examples...

264. *Comment.* What is mind-blowing (★★★) is that this characterization holds true without assuming the 2π-periodicity of the function! The true characterization of the sine function by H. DELANGE (1967) is as follows:

The sine function is the only function $f \in \mathcal{C}^\infty(\mathbb{R}, \mathbb{R})$ satisfying (1) and (2).

It is disconcerting that these sole imposed assumptions on f force its 2π-periodicity. The proof of this characterization is however more difficult than in the simplified context presented above.

266. *Comment.* A generalization of the above result to $\mathbb{R}^n$, with a coordinate system $(Ox_1 x_2 ... x_n)$, is possible. If, for $i = 1, 2, ... , n$, $\mathcal{F}_{x_i}$ denotes the orthogonal projection of $\mathcal{F}$ on the plane with equation $x_i = 0$, one has:

$$(card \; \mathcal{F})^2 \leqslant (card \mathcal{F}_{x_1}) \times (card \mathcal{F}_{x_2}) \times ... \times (card \mathcal{F}_{x_n}). \tag{2}$$

267. *Answer.* Yes indeed, A and B are equal.

268. *Answers.* 1°) $S + T =$ **6397665634**96986126162362309531544878969887106.

$U =$ **6397665634**848672580686235832221685757584124416.

2°) $\varepsilon_r \simeq 9.47 \times 10^{-12} < 10^{-11}$.

Comments. - $S + T$ and U cannot be "mathematically equal"; if so, that would contradict the FERMAT–WILES theorem (1995), which states that there are no positive integers $x, y, z, n > 3$ satisfying $x^n + y^n = z^n$.

- This example is due to D.X. COHEN, in Simpson Horror Show VI (1995). It is optimal, in terms of ε_r, if one considers $S < 5000$, $S < T < 1.5 \times S$ and $3 < k < 50$; if one looks further, up to $S = 10000$, one can do slightly better, *i.e.* get a smaller ε_r (V. FEUVRIER, personal communication, 2015).

269. *Comments.* In the same vein, we have the following interesting results:

- there are infinitely many cases where squares of integers equal cubes of integers, an example is $8^2 = 4^3$;

- there is only one case where a squared integer is immediately followed by a cubed integer: $8 = 2^3$ followed by $9 = 3^2$;

- (more general): the only integers $m \geqslant n \geqslant 2$ such that $m^n - n^m = 1$ are $m = 3$ and $n = 2$. This is known as the theorem of MORET-BLANC (1876);

- (even more general): the only integers a, b, c, d, all $\geqslant 2$, satisfying $a^b - c^d = 1$ are $a = 3, b = 2, c = 2, d = 3$. This result, conjectured by E. CATALAN (1844), has finally been proved by P. MIHAILESCU in 2002.

270. *Answers.* The answer is Yes to both questions.

First question. If $\left(\sqrt{2}\right)^{\sqrt{2}}$ is rational, then we have answered the question (the irrational $\sqrt{2}$ raised to the irrational $\sqrt{2}$). If $\left(\sqrt{2}\right)^{\sqrt{2}}$ is irrational, then $\left(\left(\sqrt{2}\right)^{\sqrt{2}}\right)^{\sqrt{2}} = 2$ is an example which answers the question.

Second question. If $\left(\sqrt{2}\right)^{\sqrt{2}}$ is irrational, then we have answered the question (the irrational $\sqrt{2}$ raised to the irrational $\sqrt{2}$). If $\left(\sqrt{2}\right)^{\sqrt{2}}$ is rational, then $\left(\sqrt{2}\right)^{\sqrt{2}} \times \sqrt{2}$ is irrational (product of a rational by an irrational); hence we have answered the question since $\left(\sqrt{2}\right)^{\sqrt{2}} \times \sqrt{2} = \left(\sqrt{2}\right)^{\sqrt{2}+1}$ (the irrational $\sqrt{2}$ raised to the irrational $\sqrt{2} + 1$).

Comment. Actually, $\left(\sqrt{2}\right)^{\sqrt{2}}$ turns out to be irrational...

271. *Answer.* We necessarily have $u_n = n$ for all $n \in \mathbb{N}$.

148

Comment. A result in the same vein, even easier to prove, is as follows. Suppose that, instead of (1), we have

$$u_n + u_{u_n} + u_{u_{u_n}} = 3n \text{ for all } n \in \mathbb{N}. \tag{2}$$

Then, again, we necessarily have $u_n = n$ for all $n \in \mathbb{N}$.

273. *Answers.* 1°) We have: $D_1 = 1, D_2 = 2, D_3 = 5$.

3°) One proves by induction that $D_n \leqslant n!$ for all n. Consequently, $\left| \frac{D_k}{k!} x^k \right| \leqslant |x|^k$, which ensures that the radius of convergence R of the power series with general term $\frac{D_k}{k!} x^k$ is greater than 1. Working for $|x| < R$, one gets

$$f'(x) = \sum_{n=1}^{+\infty} \frac{D_n}{(n-1)!} x^{n-1} = \sum_{n=0}^{+\infty} \left(\sum_{k=0}^{n} \frac{1}{(n-k)!} \frac{D_k}{k!} \right) x^n,$$

which is the CAUCHY product of the power series $\sum_{n=0}^{+\infty} \frac{x^n}{n!}$ and $\sum_{n=0}^{+\infty} \frac{D_n}{n!} x^n$. Thus, $f'(x) = e^x f(x)$. Finally, the sum f is the solution of the CAUCHY problem

$$\begin{cases} f'(x) = e^x f(x) \\ \quad f(0) = 1. \end{cases}$$

Consequently,

$$f(x) = e^{(e^x - 1)} \text{ for } |x| < R.$$

274. *Answer.* 2°) Let $f : [0, 1] \to [0, 1]$ be defined as follows:

$$f(x) = \begin{cases} 3/4 & \text{if } 0 \leqslant x \leqslant 1/4, \\ -3x + (3/2) & \text{if } 1/4 \leqslant x \leqslant 1/2, \\ 0 & \text{if } 1/2 \leqslant x \leqslant 1. \end{cases}$$

The only fixed point of f is $3/8$, while $x_0 = 1/4$ yields $x_1 = 1/2, x_2 = 1/4$, etc.

Comment. M. KRASNOSELSKI (1920–1997) was a Soviet mathematician renowned for his works in Analysis and its applications.

275. *Comment.* L. FEJER (1880–1959) was a Hungarian born mathematician. He was the thesis advisor of brilliant mathematicians such as J. VON NEUMANN, P. ERDÖS and G. POLYA.

276. *Answers.* 1°) $f(x) = \pm 1$; $f(x) = \cos(x)$; $f(x) = \cosh(x)$.

2°) Necessarily $f'(0) = 0$; $f(0) = \pm 1$; $f''(x) - \frac{f''(0)}{f(0)} f(x) = 0$.

3°) The solutions of (1) are: $x \in \mathbb{R} \mapsto f(x) = \pm 1$; $f(x) = \pm \cos(kx)$ with $k \neq 0$; $f(x) = \pm \cosh(kx)$ with $k \neq 0$.

Comment. Some easier or more classical functional equations can be found in Tapas 106, 107, 108, 138, 139.

277. *Answers.* 1°) $f(x) = \exp(x^2)$.

4°) (b). $f(x) = \exp(-ax^2)$, where $a > 0$.

Comments. - The final result could be proved without following the path 1°) to 4°).

- The result of this tapa serves in characterizing the normal distribution in Probability theory.

278. *Answer.* 2°) The problem $(\mathcal{C}_3)$ has a unique solution θ_b, which is defined on the whole $\mathbb{R}$. Here it is:

- If $|b| \leqslant 1$, $\theta_b(t) = -\cosh(t) + b\sinh(t)$.

- If $b > 1$,

$$\theta_b(t) = \begin{cases} -\cosh(t) + b\sinh(t) & \text{if } t \leqslant \arg\tanh(1/b), \\ \sqrt{b^2-1} \times \sin(t - \arg\tanh(1/b)) & \text{if } \arg\tanh(1/b) \leqslant t \leqslant \arg\tanh(1/b) + \pi; \\ \theta_b(2\arg\tanh(1/b) + \pi - t) & \text{if } \arg\tanh(1/b) + \pi \leqslant t. \end{cases}$$

- If $b < -1$, take the "symmetrized" version $\theta_{-b}(-t)$ of the function appearing in the case just above.

282. *Comments.* (a) The equivalence shown above is due to C.H. ROWE (1920).

(b) A derivative function satisfies the intermediate value property. This is DARBOUX's theorem (see Tapa 123). So, what is missing for a derivative function to be continuous is precisely the closedness property (b) presented in (2).

283. *Answers.* 2°) By 0 in the first example, by $\frac{2}{\pi}$ in the second one, by $\frac{1}{2}$ in the third one.

Comment. The result of this tapa is due to D. SAADA (2009).

284. *Answer.* One possible proof is based on the next two ingredients:

- Firstly, a function admitting right-derivatives and left-derivatives at all points is continuous and differentiable except on a countable set (this is a somewhat subtle result, proved in some advanced courses on Differential calculus).

- Secondly, a mean value theorem like the following one. Let $f : \mathbb{R} \to \mathbb{R}$ be continuous and possess a right-derivative at each point of $[a, b]$; we suppose moreover that there exists a constant M such that

$$\left| f'_+(x) \right| \leqslant M \text{ for all } x \in [a, b] \setminus D, \text{ where } D \text{ is a countable set.} \tag{2}$$

Then

$$|f(b) - f(a)| \leqslant M(b - a). \tag{3}$$

150

Now, here is the proof itself. At a point x where f is differentiable, the hemi-differential relation (1) implies that $f'(x) = 0$. Furthermore, except on a countable set D, the function f is differentiable; consequently:

$$\text{For all } x \notin D, \ f'(x) = 0.$$

We then conclude with (3) by considering $M > 0$ as small as desired.

Comments. - The constant 2 is not mandatory in the hemi-differential equation (1). The result (that f is constant) still holds true if one considers the hemi-differential equation $f'_+ = kf'_-(x)$ with $k \neq 1$.

- Even more generally, to secure that f is constant on $\mathbb{R}$, it would suffice to have

$$f'_+(x) = k(x)f'_-(x) \text{ for all } x \in \mathbb{R},$$

with $k(x) \neq 1$ except on a countable set of x.

286. *Answer.* 1°) $\Delta_1(x) = 0$; if $P(x) = ax^2 + bx + c$, then $\Delta_2(x) = b^2 - 4ac$.

287. *Answers.* 1°) Only (general) quadratic functions, $f(x) = ax^2 + bx + c$, satisfy the property (1) [that therefore includes affine functions, corresponding to $a = 0$]. If we impose that $c = \frac{x+y}{2}$ is the *only* point in (a, b) for which (1) holds true, then either f or $-f$ is quadratic strictly convex [corresponding to $a \neq 0$].

2°) Surprisingly enough, in the present case, only affine functions, $f(x) = ax + b$, satisfy property (2). That means that if we impose that $c = \alpha x + \beta y$ is the only point in (a, b) for which (2) holds true, there is no function answering our question...

288. *Comments.* - Except for some special situations (like polynomial functions f), the proofs are not "constructive" in the sense that, given f, one cannot explicitly exhibit such g and h. Moreover, several diff-convex decompositions of f are possible; the meaning of "best decomposition" depends on the goal one has set.

- One cannot lower the degree of regularity of f. Let, for example, the function f of the real variable be defined by $f(x) = -|x|^{3/2}$. Then, f is not of class $\mathcal{C}^2$ and for any function h of class $\mathcal{C}^\infty$, $g = f + h$ cannot be convex.

290. *Comment.* An example of a sequence (α_n) satisfying the two required assumptions in (1) is $\alpha_n = n$. In such a case, the result of this tapa is known as CROFT's lemma.

291. *Answer.* This limit is a quotient of integrals: $\frac{\int_a^b [f(t)]^2 \, dt}{\int_a^b f(t) \, dt}$.

294. *Comments.* - These results are known as H. JORIS' theorem (1982).

- In the same vein, one could ask directly for the differentiability or not of $\sqrt{f}$. Let $f : \mathbb{R} \to [0, +\infty)$ be a function of class $\mathcal{C}^2$. Is $\sqrt{f}$ a differentiable function on $\mathbb{R}$? The answer is not obvious, think of the function: $f(x) = x^2$ for which $\sqrt{f}(x) = |x|$, and of the function $f(x) = x^4$ for which $\sqrt{f}(x) = x^2$. The trouble comes from the points where f vanishes.

Let a be a point where $f(a) = 0$. The following results may be obtained immediately: $f'(a) = 0$ necessarily, $\left(\sqrt{f}\right)'(a) = 0$ if $\sqrt{f}$ is differentiable at a. The crucial point remains the differentiability or not of $\sqrt{f}$ at a. In that respect, one can prove the following interesting result: $\sqrt{f}$ is differentiable at a if and only if $f''(a) = 0$.

296. *Comment.* This is an interesting result linking Linear algebra and Topology.

299. *Comment.* Simple counterexamples with $n = 2$ show that the (assumed) commutativity property of the matrices $A_{i,j}$ cannot be completely removed.

300. *Answers.* For $\theta \neq 0$, the eigenvalues of $M(\theta)$ are $1 - \theta$ and $1 + \theta$. The eigenspace associated with $1 - \theta$ is the vector line of $\mathbb{R}^2$ generated by $u(\theta) = \left(\sin\left(\frac{1}{\theta}\right), \cos\left(\frac{1}{\theta}\right)\right)$; the eigenspace associated with $1 + \theta$ is the vector line of $\mathbb{R}^2$ generated by $v(\theta) = \left(\cos\left(\frac{1}{\theta}\right), -\sin\left(\frac{1}{\theta}\right)\right)$.

The case of $N(\theta)$ is in the same vein as $M(\theta)$, with more "regular" entries, however. For $\theta \neq 0$, the eigenvalues of $N(\theta)$ are e^{-1/θ^2} and $-e^{-1/\theta^2}$. The eigenspace associated with e^{-1/θ^2} is the vector line of $\mathbb{R}^2$ generated by $\left(1 + \cos\left(\frac{2}{\theta}\right), \sin\left(\frac{2}{\theta}\right)\right)$; the eigenspace associated with $-e^{-1/\theta^2}$ is the vector line of $\mathbb{R}^2$ generated by $\left(1 - \cos\left(\frac{2}{\theta}\right), -\sin\left(\frac{2}{\theta}\right)\right)$.

These two examples show that, while the eigenvalues are (here) continuous functions of the parameter θ, the unit vectors directing the eigenspaces behave very badly when $\theta \to 0$.

301. *Answer.* M is diagonalizable if and only if a_i and a_{n+1-i} are either both null or both non-null, and this holds for all integers $i < \frac{n+1}{2}$.

302. *Comment.* We do not know for which integers $n \geqslant 4$ we have $\pi(n) = 3$.

303. *Answers.* 1°) The answer is Yes.

2°) The answer is No. Here is a counterexample. Let $S = \left\{ \begin{bmatrix} r & 1 \\ 1 & 1/r \end{bmatrix} : r > 0 \right\}$; then S is bounded from below by the null matrix 0; one can easily prove that the greatest lower bound or infimum of S (if any) should be the 0 matrix; but $M = \begin{bmatrix} 0 & 1 \\ 1 & 0 \end{bmatrix}$ is a lower bound for S (easy to check) and the expected inequality $M \succcurlyeq 0$ is not true.

3°) No. 4°) Yes.

305. *Comment.* The classical case of the CAUCHY–SCHWARZ inequality in $\mathbb{R}^n$ is when $A = M = I_n$.

306. *Answers.* 1°) There are three classes of matrices answering the question:

- positive definite matrices; negative definite matrices;

- indefinite matrices whose eigenvalues λ_k enjoy the following property: the absolute value of any negative eigenvalue λ_i is greater than or equal to any positive eigenvalue λ_j (*i.e.*, $|\lambda_i| \geqslant \lambda_j$).

2°) Besides positive definite and negative definite matrices, there are further matrices A answering the question, those satisfying the following property: A has only two eigenvalues, one the negative of the other; thus A is such that A^2 is a scalar matrix.

307. *Answer.* For the second question, answer by writing

$$A = (AS)(S^{-1}),$$

where S is proposed in (1). Thus, AS is symmetric since, according to (1), $(AS)^T = SA^T = AS$.

Comment. While the sum of two symmetric matrices is symmetric, it is a bit surprising that *any* matrix could be written as a product of two symmetric matrices.

309. *Comments.* These characterizations work only for symmetric matrices. For more on this (history, references, proofs,...), see

J.-B. HIRIART-URRUTY, *Le théorème déterminantal en bref.* Revue de la Filière Mathématiques RMS (ex-Revue de Mathématiques Spéciales), n° 119, 15–22 (2008–2009).

Equivalences [(1) ⟺ (3)] or [(1) ⟺ (2)] are proved in the references below:

H. CARRIEU and P. LASSÈRE, *One more simple proof of the* CRAIG–SAKAMOTO *theorem.* Linear Algebra Appl. 431, 1616–1619 (2009).

X. BONNEFOND, *An analytical proof of* OGAWA*'s determinantal theorem.* Aequat. Math. 83 (2012), 127–130.

311. *Comment.* An extension of the result to the product of three positive definite matrices is possible, but not to the product of four positive definite matrices.

312. *Comment.* The *quadratic* character of the objective function f to be minimized as well as the *polyhedral* character of the constraint set C play an important role here. To assume that f is convex, which amounts to requiring that A is positive semidefinite, simplifies the proof a bit, but not so much.

The existence result stated in this tapa is due to FRANK and WOLFE (1956).

313. *Comment.* The norms ν_1, ν_2, as well as

$$M \mapsto \nu_3(M) = \sigma_1(M) + \sigma_2(M) + \dots + \sigma_n(M),$$

are norms on $\mathcal{M}_n(\mathbb{R})$, corresponding to the usual three norms $\|\cdot\|_2$, $\|\cdot\|_\infty$ and $\|\cdot\|_1$ on $\mathbb{R}^n$.

314. *Answers.* 1°) The answer is Yes. To prove it, decompose P as a product of irreducible polynomial factors either of degree 1 (in that case the exponent of this factor has to be even) or of degree 2 (in that case, since the discriminant is negative, the canonical decomposition provides a sum of squares); then note that a product of sums of squares is a sum of squares.

2°)−3°). Both can be treated in the same manner. Let $q(x) = x$ or $1 - x^2$; one posits

$$\mathcal{E}_q = \big\{\text{polynomial functions } P \text{ decomposable as } A^2 + q(x)B^2\big\}.$$

A key-point is to show that $\mathcal{E}_q$ is stable under multiplication. Then, prove the announced statements for polynomial functions of degree at most 2. Finally, use decompositions of polynomials as products of polynomials of degree at most 2.

315. *Comments.* - For MARDEN's theorem, see the comments in Tapa 215.

- Unfortunately, there is no MARDEN type theorem for arbitrary quadrilaterals.

316. *Comment.* A strange result indeed... but periodic functions can be bumpy.

317. *Comments.* - This result was firstly proved by JÖRGENS (1954). Since $\mathbb{R}^2$ can be identified with $\mathbb{C}$, techniques and results from the field of Complex analysis could also be used.

Extensions to some other cases than functions of two variables ($n = 2$), like for $n = 3$ or $n = 5$, were made by CALABI and POGORELOV. The general result, for functions of n variables, is due to POGORELOV.

- The announced result does not hold true if we just assume that $\det\left[\nabla^2 f(x,y)\right] = 1$ for all (x,y) in some open subset of $\mathbb{R}^2$.

318. *Answer.* Surprisingly enough, f has to be an affine function. There are several ways of proving this (using differential calculus, relying on existence results for differential equations, etc).

Comment. The announced result does not hold true if we just assume that $\|\nabla f(x)\| = 1$ for all x in some open subset of $\mathbb{R}^n$.

319. *Comment.* This example was given by T. MOTZKIN in 1967.

320. *Answer.* In the first question, K is the unit circle of $\mathbb{R}^2$, which is not convex.

Comment. The convexity of K, for $n \geqslant 3$, is due to L. BRICKMAN (1961). For a rather simple proof, see

H. PÉPIN, Revue de la Filière Mathématiques RMS (ex-Revue de Mathématiques Spéciales), 3 (2004), pp. 171–172. Answer to a problem posed by J.-B. HIRIART-URRUTY.

321. *Answer.* The answer is Yes. To prove that a convex combination of two points x and y in C still lies in C, connect x and y with a continuous map, and use the intermediate value property for continuous functions.

Comments. - Note that property (1) is satisfied by any discrete set C in $\mathbb{R}^n$.

- According to the result proved in this tapa, among the closed arcwise-connected sets C in $\mathbb{R}^n$, those which are convex are exactly those satisfying the property (1).

322. *Answer.* The answer is Yes.

323. *Answers.* 1°) The answer is Yes. Use resources from Linear algebra to prove it.

2°) The answer is No. A counterexample is given by the following convex compact polyhedron $\mathcal{A}$ of $\mathcal{M}_2(\mathbb{R})$:

$$\left\{ \begin{array}{c} \mathcal{A} \text{ is the convex hull of the four matrices} \\ \begin{bmatrix} 1 & 0 \\ -2 & -1 \end{bmatrix}, \begin{bmatrix} 1 & 0 \\ 2 & -1 \end{bmatrix}, \begin{bmatrix} -1 & -2 \\ 0 & 1 \end{bmatrix}, \begin{bmatrix} -1 & 2 \\ 0 & 1 \end{bmatrix} \end{array} \right\},$$

and B is chosen as $\begin{bmatrix} 1 & 0 \\ 0 & 1 \end{bmatrix}$.

324. *Answers.* 1°) Let y_0 be a non-null element of $\mathbb{R}^m$, and let $f : \mathbb{R}^n \to \mathbb{R}^m$ be defined as follows: $f(0) = 0, f(x) = y_0$ if $x \neq 0$.

2°) A proposal: $f(x) = \sin\left(\frac{1}{x}\right)$ if $x \neq 0, f(0) = 0$.

326. *Answers.* 1°) The spherical loaf of bread is obtained by rotating the curve $y = f(x) = \sqrt{r^2 - x^2}$ around the x-axis. We therefore have, for a slice with cuts at a and $a + h$:

$$V = \pi \int_a^{a+h} [f(x)]^2 \, dx = \pi r^2 h - \frac{\pi}{3} h(h^2 + ah + a^2). \tag{1}$$

With $h = 2r/n$, as expected, this volume is maximal when the cutting abscissa is around 0.

2°) For a slice with cuts at a and $a + h$, the area of the crust is:

$$A = 2\pi \int_a^{a+h} f(x)\sqrt{1 + [f'(x)]^2} dx = 2\pi r h. \tag{2}$$

This is independent of the cutting abscissa a! Thus, all slices of thickness h have the same surface area A, *i.e.*, all slices of bread (in the middle of the loaf or close to its edges) have exactly the same area of crust! With $h = 2r/n$, as expected we get that $A = (4\pi r^2)/n$ (the total crust of the spherical loaf divided by the number of pieces n).

155

327. *Answers.* 1°) We have: $V_1 = 2, V_2 = \pi, V_3 = \frac{4\pi}{3}, V_4 = \frac{\pi^2}{2}, V_5 = \frac{8\pi^2}{15} \simeq 5.2638, V_6 = \frac{\pi^3}{6}$.

A general formula for V_n is: $V_n = \frac{\pi^{\frac{n}{2}}}{\Gamma(\frac{n}{2}+1)}$.

Since $V_{n+2}/V_n = \pi/(\frac{n}{2}+1)$, it follows that $V_{n+2} < V_n$ for $n \geqslant 5$ and $V_{n+2} > V_n$ for $n \leqslant 5$. Thus V_5 is the maximal volume.

2°) Expression for the sum of the series: $\sum\limits_{n=1}^{+\infty} V_n = e^\pi \left(1 + \frac{2}{\sqrt{\pi}} \int_0^{\sqrt{\pi}} e^{-t^2}\, dt\right) - 1$. To obtain this, separate the terms of even n (whose sum determines an exponential function) and the terms of odd n (whose sum can be recognized as the solution of a differential equation $y'(x) = 1 + xy(x)$).

3°) We have that $S_n = nV_n$. Thus $S_{n+2}/S_n = 2\pi/n$. Consequently, S_n is maximal for $n = 7$. The value of S_7 is $16\pi^3/15 \simeq 33.073$.

Comments. The occurrence of the "maximizing" integers $n = 5$ (for volumes) and $n = 7$ (for surface areas) is a bit strange...

For more on these questions, see

J.-B. HIRIART-URRUTY and M. PRADEL, *Les boules !* Quadrature, n° 54, 8–11 (2004).

328. *Comments.* - The result and inequality (1) in the first question are due to EULER.

- The result and inequality (2) of the second question are due to ERDÖS and MORDELL.

- There is another result, similar to (2) but easier to obtain, which is due to G. STEENSHOLT (1956). Here it is. Let ABC be a triangle with sides BC, CA and AB (of lengths a, b, c, respectively). As above, for an interior point P of ABC, let its distances from A, B and C be R_1, R_2, R_3, and let its distances from the sides BC, CA and AB be r_1, r_2, r_3. Then:

$$aR_1 + bR_2 + cR_3 \geqslant 2(ar_1 + br_2 + cr_3). \tag{3}$$

The idea of the proof is to play with the (expressions of the) areas of the triangles PBC, PAC, PAB, whose sum is the area of ABC. In the case of an equilateral triangle, $a = b = c$, this gives the original ERDÖS–MORDELL inequality (2).

329. *Comment.* A generalization of the result to a convex compact polyhedron S in $\mathbb{R}^n$ is possible.

330. *Answers.* 1°) Since the derivation operation is linear, it is continuous on $\mathcal{C}^\infty(\mathbb{R})$ if and only if there exists a constant L such that

$$\|f'\| \leqslant L \|f\| \text{ for all } f \in \mathcal{C}^\infty(\mathbb{R}). \tag{1}$$

Choose $f(x) = \exp(nx)$, with n a positive integer, to see that an inequality like (1) cannot be secured for all n.

2°) For the vector space $\mathcal{P}$ of real polynomial functions, the situation is different: there are norms on $\mathcal{P}$ for which the derivation operation is continuous, and there are norms for which the derivation operation is not continuous. The derivation operation is continuous on $\mathcal{P}$ (equipped with $\|\cdot\|$) if and only if there exists a constant K such that

$$\|P'\| \leqslant K \|P\| \text{ for all } P \in \mathcal{P}. \tag{2}$$

As for examples:

- if we equip $\mathcal{P}$ with the norm $P \mapsto \|P\|_1 = \sum_{k=0}^{+\infty} |P^{(k)}(0)|$ (actually a finite sum), the derivation operation is continuous ((2) holds true with $K = 1$);

- if we equip $\mathcal{P}$ with the norm $P \mapsto \|P\|_2 = \max_{x \in [-1,1]} |P(x)|$, the derivation operation is not continuous (consider $P(x) = x^n$ to see that (2) cannot be secured).

331. *Answers.* 2°) We have:

$$a_0 = \frac{1}{2\pi}(e^{2\pi} - 1); a_n = \frac{1}{\pi(n^2 + 1)}(e^{2\pi} - 1) \text{ (for } n \geqslant 1).$$

$$b_n = \frac{-n}{\pi(n^2 + 1)}(e^{2\pi} - 1) \text{ for } n \geqslant 1.$$

4°)–5°) We get:

$$\sum_{n=1}^{+\infty} \frac{\cos n}{n^2 + 1} = \frac{1}{2}\left[\frac{e + e^{2\pi - 1}}{e^{2\pi} - 1}\pi - 1\right]; \tag{2}$$

$$\sum_{n=1}^{+\infty} \frac{1}{n^2 + 1} = \frac{1}{2}\left[\frac{e^{2\pi} + 1}{e^{2\pi} - 1}\pi - 1\right]. \tag{3}$$

Let us admit that, without the magic of FOURIER series, it would have been more difficult to obtain results (2) and (3)...

332. *Comment.* There are other similar characterizations of the sine function; here are some of them:

- The sine function is the only one minimizing

$$\mathcal{I}_2(f) = \int_0^{\frac{\pi}{2}} \left([f'(x)]^2 - [f(x)]^2\right) \, dx$$

among the set of functions $f \in \mathcal{C}^1\left([0, \frac{\pi}{2}], \mathbb{R}\right)$ satisfying

$$f(0) = 0, \ f'(0) = 1.$$

- The sine function is the only one minimizing

$$\mathcal{I}_3(f) = \int_0^{2\pi} \left([f'(x)]^2 - [f(x)]^2\right) \, dx$$

157

among the set of functions $f \in \mathcal{C}^1([0, 2\pi], \mathbb{R})$ satisfying

$$f(0) = 0, \ f'(0) = 1, \ \int_0^{2\pi} f(x) \, dx = 0.$$

- The sine function is the only one minimizing

$$\mathcal{I}_4(f) = \int_0^\pi \left([f''(x)]^2 - [f(x)]^2 \right) \, dx$$

among the set of functions $f \in \mathcal{C}^2([0, \pi], \mathbb{R})$ satisfying

$$f(0) = 0, \ f'(0) = 1, \ f(\pi) = 0.$$

For more on the question of characterizing the sine function (various proposals, references, etc.), see

J.-B. HIRIART-URRUTY, *Sinusite : diagnostic, caratérisations, traitements....* Quadrature, n° 99, 10–15 (2016).

333. *Answer.* 2°) We have that:

$$\max \left\{ \sum_{i=1}^n f(a_{ii}) : A = [a_{ij}] \in \mathcal{E}(M) \right\} = \sum_{j=1}^n f(\lambda_j).$$

158

References and sources

Section "Problems and solutions" of *The American Mathematical Monthly*
http://www.maa.org/publications/periodicals/american-mathematical-monthly

Rubrique "Questions et réponses" de la *Revue de la Filière Mathématique*
(ex-Revue de Mathématiques Spéciales, RMS in short)
http://www.rms-math.com/

Revue *Quadrature*
http://www.quadrature.info/